DEVELOPMENT POLICY AND PLANNING

Economic controls are giving way to market forces as the main vehicle of development. Yet the techniques and models of development planning are still relevant to policy makers: governments must continue to exercise economic management and direction to accelerate market-led growth.

Development planning provides a useful set of analytical techniques to facilitate economic decision making and management and, by assembling a set of empirical relationships for the whole economy, identifies the effects of particular policy changes. With the continuing evolution of techniques, models can now be designed to match the constraints and objectives of individual economies.

Development Policy and Planning provides a lucid explanation of the main techniques used for economic policy formulation in developing countries, and discusses their application. Including growth modelling, social accounting matrices, computable general equilibrium models, linear programming and cost–benefit analysis, this book describes each technique and, by the use of empirical examples, illustrates the methods by which economies can be guided towards their development goals.

Anis Chowdhury is Senior Lecturer in Economics at the University of Western Sydney. **Colin Kirkpatrick** is Professor of Development Economics at the University of Bradford.

DEVELOPMENT POLICY AND PLANNING

An introduction to models and techniques

Anis Chowdhury and
Colin Kirkpatrick

London and New York

First published 1994
by Routledge
11 New Fetter Lane, London EC4P 4EE

Simultaneously published in the USA and Canada
by Routledge
29 West 35th Street, New York, NY 10001

© 1994 Anis Chowdhury and Colin Kirkpatrick

Typeset in Garamond by
Mathematical Composition Setters Ltd, Salisbury, Wiltshire

Printed and bound in Great Britain by
Mackays of Chatham PLC, Chatham, Kent

British Library Cataloguing in Publication Data

A catalogue record for this book is available from the British Library

ISBN 0-415-09888-2
ISBN 0-415-09889-0 Pbk

Library of Congress Cataloging in Publication Data

Chowdhury, Anis, 1954–
    Development policy and planning: an introduction to models and
techniques / Anis Chowdhury and Colin Kirkpatrick.
        p. cm.
    Includes bibliographical references and index.
    ISBN 0-415-09888-2 : $45.00 (U.S.)—ISBN 0-415-09889-0 (pbk.) :
$14.95 (U.S.)
    1. Economic development—Mathematical models.   2. Economic
policy—Mathematical models.   3. Planning—Developing countries—
Mathematical models.   I. Kirkpatrick, C.   II. Title.
HD75.5.C52   1994
338.9′001′5118—dc20   93-7404
                                                                CIP

To Our Families

CONTENTS

vii

CONTENTS

FIGURES

TABLES

x

PREFACE

The 1980s was a decade in which many developing countries adopted significant changes in economic policy, with a reorientation towards a market-based approach to development policy in which the private sector plays a major role. This contrasts sharply with the earlier post-independence period, when development planning was widely adopted throughout the Third World. So where does development planning now stand as part of development policy management?

This book has arisen out of the authors' conviction that the techniques and models of development planning are as relevant to policy makers who are now involved in 'development policy management' as they were to those pursuing 'development planning' in an earlier period. Although economic controls have given way to market forces as the main vehicle of development, governments must continue to exercise a degree of economic management and direction which will accelerate the growth process beyond what might be achieved by market forces alone.

Development planning fulfils several functions. By providing an analytical framework of the economy's structure, it allows the policy maker to evaluate the prospects for, and constraints upon, economic growth and structural change. By assembling a consistent and integrated set of relationships for the whole economy, it enables the policy maker to identify the direct and indirect effects of particular policy changes.

The various techniques available to policy makers are described in this book, and their application to particular policy areas is discussed. The techniques of development planning have evolved through time and, as a result, there is now a much wider range of procedures available for policy analysis. Models can more easily be designed to match the constraints and objectives of individual economies, rather than using a standard framework. Also, the development of techniques for simulating market outcomes means that policy analysis can shift away from the setting of targets to the comparison of instruments and programmes. Development planning therefore provides a useful set of analytical techniques for economic decision making and policy formulation, which continue to be widely used in developing countries.

Much of the published material on development planning methods is written at a relatively advanced level, and is often presented in mathematical terms. The authors' objective is to make these techniques more accessible to students of development economics, and to persons involved in economic policy formulation. It is based on the authors' experience in teaching introductory and intermediate level courses in development planning and policy to students studying the problems of Third World countries, many of whom have been involved in economic policy formulation and implementation in their own countries on completion of their studies.

The book originated in a course which the authors taught jointly at the National University of Singapore in the late 1980s. Subsequently, the material has been presented to successive cohorts of students at the University of Bradford and the University of New England, Armidale. We are grateful to these groups of students for invaluable feedback and comments on the lecture presentations of draft chapters.

The final manuscript was written in Bradford and Sydney and was typed by Jill Gulbrandsen with unfailing efficiency and good humour.

1

DEVELOPMENT POLICY ANALYSIS AND QUANTITATIVE PLANNING METHODS

DEVELOPMENT POLICY AND PLANNING: AN OVERVIEW

The role of the state in the development process is one of the oldest topics in the economics literature, and controversy continues over the relative merits of the market mechanism as opposed to state intervention. Indeed, the relationship between governments and markets is perhaps the central issue in economic development. The question, however, is not a simple choice between *laissez-faire* and state intervention, for it is self evident that in all economies the government must exercise some degree of economic management and control. The important question, therefore, is about the nature and quality, rather than the extent, of the state's intervention in the economy.

Much of the discussion of markets and governments in the context of economic development has been conducted in terms of market failure. The early development economics literature of the 1950s and 1960s identified the inadequacies of the market mechanism as the major cause of economic backwardness in developing countries, and government intervention was regarded as the means of correcting these imperfections. A variety of types of market failure were identified as inhibiting both the efficient allocation of resources and the dynamic growth process. Indivisibilities in capacity, economies of scale, monopoly and oligopoly, and externalities, all resulted in market outcomes which deviated from the perfectly competitive, allocative-efficiency, welfare-maximizing equilibrium. In addition, inadequate information on future demand, lack of infrastructural capital and high-cost input supplies could each act as constraints on private investment decisions, and thereby reduce the economy's growth rate.

1

These various forms of market failure provided a prima-facie case for government intervention in the market economies of the Third World. The dominant view among economists of that time was that the problems of market failure were particularly severe in the areas of infrastructure (roads, communications, power) and industrialization. Hence, much stress was put on government policy directed towards increasing capital accumulation in these sectors.

The form of government policy that was adopted by most developing countries was direct public sector investment. To achieve this required the government to raise the level of saving, to ensure that it had control over the use of these investible surpluses, and to allocate them to investment in areas where the physical capital constraint was most severe. But how should government decision makers tackle these tasks? What policies should they adopt?

These questions can best be addressed by establishing a framework for policy analysis appropriate to the mixed economies of the developing world. The form of policy analysis that was adopted in the earlier period was often described as 'development planning'. Development planning can be defined as 'the conscious effort of a central organization to influence, direct, and, in some cases, even control changes in the principal economic variables (e.g. GDP, consumption, investment, saving, etc.) of a certain country and region over the course of time in accordance with a predetermined set of objectives' (Todaro 1971: 1). The widespread adoption of the development planning approach to economic policy formulation in turn led to the production of national development plans. A development plan is a specific set of quantitative targets to be achieved in a given period of time, and since the early 1950s more than 300 different development plans have been formulated. Development planning can take many different forms, but it is possible to specify certain common features. A development planning exercise typically involves the use of a planning model, which specifies in quantitative terms the relationships between objectives, constraints and policy instrument variables. The model is then used to calculate a feasible or consistent solution, defined as a set of values for the policy instruments that satisfies the specified objectives and does not exceed the predetermined constraints.

The early enthusiasm for development planning was gradually supplanted by a growing sense of disillusionment, such that by the end of the 1970s many economists were talking openly of the 'failure of planning'. Having reviewed the previous 30 years' experience,

Killick (1976: 103), for example, concluded that 'medium-term development planning has in most LDCs almost entirely failed to deliver the advantages expected of it'.

This disenchantment with development planning can be related to a number of influences. First, there was mounting evidence that actual performance often fell short of the plan targets. Second, the technical limitations of the planning techniques and models being used became increasingly evident through time. Third, the dynamic growth economies of East Asia were held up as confirmation of the superiority of a non-interventionist, market-based policy stance over the interventionist, development planning approach. The example of the Asian newly industrializing countries (NICs) in turn fuelled the more general resurgence of the neoclassical paradigm in development economics, with its emphasis on the role of the price mechanism in allocating resources to their most efficient uses.

Having moved from the dirigism of the 1950s and early 1960s to the neoclassicalism of the 1970s and 1980s, the current literature takes a more qualified, middle-ground position on the role of development planning. The failure to achieve plan targets can now be seen, in part, as the inevitable result of over ambiguous and unrealistic expectations as to what could be achieved in terms of economic growth and structural transformation. In addition, the limitations of the planning techniques became evident. First, there was a concentration on investment as the single determinant of growth, with other factors such as human capital and productivity growth being ignored. Second, the models were essentially closed economies, assuming exogenously determined exports and focusing entirely on internal policies. Third, a narrow range of objectives was considered, with the targets being specified in terms of aggregate income or employment growth, which excluded consideration of poverty alleviation or income redistribution, for example. Finally, the early models typically took prices as fixed, which ruled out the use of prices as a policy instrument, and thereby made quantity-controls the main policy planning variable.

As the techniques of development planning have evolved through time, many of these limitations of the early models have been overcome. Input–output techniques allowed the planner to consider intersectoral resource allocations. Programming models brought out the implications of different constraints, including skilled labour and foreign resources. The social accounting matrix approach provided a method of modelling the effects of various policy interventions on income distribution. The development of computable general

3

equilibrium models introduced the possibilities of factor substitution and productivity increases, and allowed the planner to simulate the effects on the economy of specific policy instruments.

As a result of these developments, there is now a much wider range of techniques and procedures available for policy analysis. Models can more easily be designed to match the constraints and policy objectives of individual countries, rather than using a standard framework. Also, the shift towards simulating market outcomes means that policy analysis has shifted away from the setting of targets to the comparison of instruments and programmes.

A considered evaluation of the East Asian NICs' experience makes it clear that the neoclassical litany about 'getting the prices right' seriously distorts reality. The evidence demonstrates that the state played an active role, which went well beyond the prescribed neutral policy regime of neoclassical economics, in planning and directing the economic growth of these newly industrializing economies. What distinguishes planning in South Korea and Taiwan from that in India, for example, is the way in which the planning was market-enhancing rather than market-supplanting. Investment was directed into areas which were judged to have long-term growth potential and policy was aimed at increasing private sector involvement in targeted areas of production. Areas for intervention were selected with the objective of achieving international standards of productive efficiency and competitiveness.

The arguments against planning, therefore, are by no means conclusive. Much of the criticism amounts to a case against a particular type of development planning which was attempted in the 1950s and 1960s. Viewed as a broader process for policy analysis, development planning continues to provide a useful framework for economic decision making and policy formulation which is still widely used in developing countries.

CHARACTERISTICS OF DEVELOPMENT PLANNING MODELS

The process of planning involves the examination of a host of social and economic variables. These socio-economic variables are related to one another in a very intricate and complex manner and our understanding of the long chain of interaction becomes hazy without the aid of an analytical model. Models are needed, therefore, to analyse complex interactions between various elements which may appear to be unrelated. As Myrdal has put it:

> Models are essential aids to clear thinking ... The first virtue of models is that they can make explicit and rigorous what might otherwise remain implicit, vague and self-contradictory ... since ordinary thinking too often proceeds by fairly simple rule of thumb and uni-causal explanations, and rarely ascends to a complex system of interdependent relationships, model-thinking may serve as a kind of thought-therapy, loosening the cramped intellectual muscles, demonstrating the falsity or doubtfulness of generalizations, and suggesting the possibility of an interdependence previously excluded. The most justifiable claims for the use of economic models are the modest ones that they are cures for excessive rigidity of thought and exercises in searching for interdependent relationships.
>
> (Myrdal 1968: 1962–3).

Thus analytical planning models have a 'didactic' or 'thought-therapeutic' value. Even though models are simplified pictures of reality, they contribute to our understanding of some essential features of that reality.

Analytical planning models also have 'communication' value. Many different organizations and individual agents interact in the formulation and execution of a country's economic and social policies. Hence the ability of a planner to communicate with politicians, bureaucrats and others involved in the policy formulation process constitutes an important element in any type of planning and such communications can be enhanced by analytical planning models. A planning model specifies the relationships between the goals of the society and the instruments that are available to achieve them. By quantifying these relationships, the planners can simulate the effects of alternative policies on the societal objectives and check whether the overall plan or objectives are consistent and feasible in terms of capacity and resource constraints. The quantitative planning models therefore provide a framework within which the various agencies involved in the planning process can carry out a fruitful dialogue regarding the possibilities and trade-offs facing the nation. In short, planning models are useful precisely because they force the planners, policy makers and others involved in the planning process to set out the structure of the economy and to focus on the relationships that determine the outcome of policy changes.

A further value of analytical planning methods is that they require a comprehensive data base, which forces the economists and statisticians to assemble existing data into a consistent and accessible form, and to identify gaps where additional information is needed.

We should bear in mind, however, that exercises with analytical planning models do not provide 'the' solution. Such exercises only assist in finding them. To quote Kornai, a pioneer in model building and a practitioner of planning, 'we cannot expect our model to give final, decisive answers; it can be considered an accomplishment if it only inspires interesting thoughts, if it furnishes additional points of view for a decision' (Kornai 1975: 19). Furthermore, even though model exercises are essential elements in the preparation of well-coordinated policies, they cannot do the job all by themselves. Again, in the words of Tinbergen, another pioneer in the practice of quantitative planning techniques, 'Models constitute a framework or a skeleton and the flesh and blood will have to be added by a lot of common sense and knowledge of details' (Tinbergen 1981: 15). Therefore, the use of quantitative planning techniques cannot completely replace intuitive judgements based on experience of the working of the economic system.

Furthermore, there is no single model that can be regarded as the best. Both in theory and in practice, different types of models are suitable for examining different policy problems. Some of the model characteristics are discussed below.

Coverage

In terms of coverage or scope, planning models can be classified into: (i) overall or national models, (ii) sectoral or regional models, (iii) special models and (iv) project analysis.

The overall models deal with the entire economy and the nation's development strategy is analysed within them. The sectoral and regional models deal with individual producing sectors and regions and can be used to examine the consistency and feasibility of the overall objectives. The special models are designed for selected aspects of the overall plan, e.g. foreign trade or manpower development. Plans are ultimately implemented through projects, and it is at this stage that project analysis is used to examine the choice of techniques, location and size of plants within the overall objectives of the national plan. The technique that is most widely used for this purpose is social cost–benefit analysis.

Aggregation

Planning models can also be classified in terms of the degree of aggregation: (i) aggregate models, (ii) main-sector models and (iii) multi-sector models. The aggregate models treat the entire economy (or a region, depending on its scope or coverage) as one producing sector and are concerned with the forecasts of such major national accounts aggregates as savings, investment and gross national product (GNP). The most representative example of this type is the Harrod–Domar model.

The main-sector models divide the economy (or a region, as the case may be) into a few producing sectors and examine the interrelationships between them. Early examples of main-sector models can be found in Arthur Lewis's dual economy hypothesis which dichotomized the economy into a traditional (agricultural) and a modern (manufacturing) sector, and the Marxian schema of consumption and capital goods sectors. One consumption–capital goods version of the main-sector models, known as the Mahalanobis model, was used extensively in the early planning exercises in India.

The multi-sector models divide the economy or region into a large number of producing sectors. The core of such models is the Leontief input–output analysis. One of the advantages of the multi-sectoral models is that they provide systematic linkages between the overall and the sectoral plans. In addition, they provide a framework for consistency checks among the various sectoral plans.

Time

There are two aspects of the time dimension in planning models. The first is how far into the future a model is designed to project and the second is the way time is treated within a model. With respect to the first criterion, we can distinguish three broad types of planning models: (i) short-term, covering 1–3 years; (ii) medium-term, covering a 3–7 year horizon; and (iii) long-term, extending to 10 years or more. Most of the models that are discussed in this volume are intended for medium-term planning.

In the treatment of time, models can be either (i) static or (ii) dynamic. The static models compare one future date with the present: they indicate the future values of the model variables but do not describe the path of the economy between the starting and end periods. In contrast, the dynamic models incorporate endogenous variables from a number of time periods and provide information on the nature of movement of the economy from the present to some target year. Since the analysis of dynamic models requires a knowledge of difference or differential equations which are fairly advanced in level, these models are not discussed in this volume.

Behaviour

Planning models can be classified as either stochastic or deterministic systems, depending on the way in which behavioural relationships are treated. In the stochastic models, behavioural relationships, e.g. savings and investment, are estimated by using econometric models which allow for stochastic or random disturbances. Short-term macroeconomic models are in general stochastic. Hence results obtained from such models must be treated as probabilistic. In contrast, deterministic models, such as the open input–output system, do not specify any behavioural relations and instead treat them as exogenously or administratively determined.

Closure

Sometimes the number of equations is less than the number of variables in a model, i.e. some of the variables are not explained within the model. Thus, a modeller must choose the variables to be explained (endogenous variables) to close the model. The closure rules, i.e. which variables should become endogenous and which variables should be treated as exogenous, depend on the problems at hand. Accordingly, planning models can be classified into (i) open systems, (ii) closed or fully determined systems and (iii) partially determined systems. An example of the open system is the input–output model which treats the final demands as exogenous or given (i.e. not explained within the model). However, the final demands can be endogenized by using some form of Engel curve or demand function, linking expenditure with income. In this case, all the variables or unknowns can be calculated within the system once certain policy variables are specified, and the model then becomes 'closed', or fully determined. These models are commonly known as simulation or forecasting models. The simulation models provide alternative scenarios for different possible sets of policy or exogenous variables and thereby help evaluate outcomes of alternative policy measures.

8

However, in some cases the solution of the system is not determinate. That is, more than one value of the unknowns is possible for a given set of exogenous or policy variables. In that case, some sort of optimization technique must be used to choose the set of values which maximizes the objectives of the overall plan, given the constraints. Linear programming is the technique most often selected; hence this set of techniques is referred to as programming models.

Accounting framework

Every economic model must satisfy some definitional equality, known as an identity. This follows from the fact that all models have either an implicit or an explicit accounting framework. This is because every income must be matched by corresponding outlays as every receipt is also an expenditure. The accounting framework provides the data base for the implementation of the respective model. Hence planning models can also be classified according to their data base or accounting framework. For example, the national income accounts yield the macroeconomic identities, (e.g. GNP is the sum of final expenditures) required to close the aggregate models. Accordingly, the aggregate models can be classified as national-accounts based models. Similarly, the main-sector and multi-sector models are input–output-accounting-based models.

Models that make use of identities derived from the accounting frameworks are generally referred to as consistent models. This is because the accounting identities imply consistency between sectoral supplies and total supply and between total supply and total demand. The consistency models do not in general contain any *ex ante* welfare or optimization functions and guide resource allocation according to some prespecified goals without examining whether these are optimal outcomes. As opposed to consistency models, there are programming models which allow for inequality relationships and deal with optimum and efficient allocation of scarce resources.

It should be clear by now that the above schema of classification is not mutually exclusive. Neither is it possible to identify a single set of model characteristics which may be regarded as the best. This is because what is best for examining one type of problem may not be the most suitable for another kind of policy dilemma. Therefore, the choice of a model must be guided by the problems at hand. The planner usually deals with a variety of interrelated problems and hence must use more than one planning model. In fact, as the classification schema shows, the various planning models have overlapping components. This implies that the results obtained from one type of model can be fed into another. This in turn forces the estimates from different model exercises to be consistent. The interrelated nature of various types of planning models becomes clear if we view the planning process as 'planning in stages'. In this regard, it is relevant to quote Tinbergen and Bos (1962: 10):

> the first stage may consist of a macro-economic study of the general process of production and investment, along the same lines suggested by Harrod–Domar models or by similar, somewhat more complicated models. The aim of this first stage should be to determine, in a provisional way, the rate of savings and the general index of production. A second stage may consist then in specifying production targets for a number of sectors over a fairly long period. A third stage, if needed, may go into more detail for a shorter period, giving figures for a larger number of smaller sectors. A fourth stage may consist in 'filling the plan out' with individual projects. Intermixed with this succession there may be stages of revision of the previous stages. Thus, the figures of the second stage may already enable the planner to revise some of the coefficients used in the first stage and to re-do therefore, the first stage. After a fixed interval of time, new data will be available and this may lead to another revision, combined or not with shifting the period of the plan.

In the subsequent chapters, we shall discuss the Harrod–Domar model, the two-gap model, the Feldman–Domar–Mahalanobis model, the input–output model and the social-accounting-matrix-based models as illustrations of planning techniques that are used in some form or other in developing countries. As mentioned earlier, we shall focus only on the static versions of these models. In terms of the above schema of classification, they fall under various categories. While all of them are economy-wide consistency models, the Harrod–Domar and the two-gap models are aggregate and the rest are disaggregated in nature. They are deterministic and most do not involve any optimizing criteria. One popular version of the programming models – the linear programming technique – used for optimizing some social objective function is discussed in this volume. One of the limitations of these models is that the price mechanism does not play any role in them and hence they fail to reflect the interdependence in a decentralized decision-making setting. As a response to this shortcoming, multi-sectoral planning models are extended to computable general equilibrium models (CGE) and CGE techniques are being used increasingly to analyse policy problems in developing countries. The discussion on planning models therefore includes a chapter on CGE. The final chapter is concerned with cost–benefit analysis. This is the most disaggregated planning technique discussed in this book, and deals with the appraisal of individual investments as a means of selecting between alternative projects.

FURTHER READING

Todaro (1989: ch. 16) gives a good introduction to the theory and practice of development planning, while the experience with development planning is examined from a neoclassical perspective in Little (1982: ch. 3). Gillis *et al.* (1992: ch. 6) summarize different types of planning models. Blitzer *et. al.* (1975) is a more advanced text: Chapter I has a clear discussion on different types of planning models and Chapters II and III discuss uses of the models.

2

AGGREGATE CONSISTENCY MODELS

The process of economic growth is central to the economics of development, and growth models represent the earliest approach to development planning. In this chapter we shall examine two simple and well-known aggregate consistency models: the Harrod–Domar (H–D) model and the two-gap model.

THE HARROD–DOMAR GROWTH MODEL

In drawing up a development plan, the usual first step is to assess the aggregate rate of growth of the economy that can be attained within the prevailing socio-economic conditions. The possible overall rate of growth can most easily be determined by analysing the past and present relationships between such aggregate variables as gross domestic product (GDP), consumption, savings, investment and the rate of population growth. The H–D growth model provides the simplest possible framework within which the relationships among the aggregate macro variables can be examined. Despite its simplicity, a host of planning problems and a wide range of possibilities can be analysed within the H–D framework. In fact, the H–D model or some variant of it is the most widely used quantitative planning technique and, even though many plan documents do not explicitly present the H–D model, elements of it can be found in the way investment requirements and the role of savings are analysed in the formulation of the economic growth plan.

The basic H–D model can be summarized as follows. We make the following assumptions.

1 Savings S is a simple proportional function of national income Y. That is,

$$S = sY \tag{2.1}$$

where s is the average propensity to save.

12

2 The amounts of capital K and labour L required to produce any given flow of output Y are 'uniquely' given. Thus, the aggregate production function implied by the H–D model can be expressed as

$$Y = \min(K/v, L/u) \qquad (2.2)$$

where $u = L/Y$ is the amount of labour required to produce one unit of output, or the reciprocal of labour productivity, and $v = K/Y$ is the amount of capital required to produce one unit of output, or the reciprocal of capital productivity.

The implication of this type of 'fixed proportion' production function is that the output will be determined by the available quantity of labour and capital, whichever is the lesser. Since the developing countries are usually labour surplus (relative to capital) economies, it follows that capital is the determining factor for the growth of output. This conclusion can be derived formally as follows.

From the national income equilibrium conditions we know that

$$S = I \qquad (2.3)$$

where I is the net investment (assuming no depreciation, for simplicity). Also, by definition,

$$I = \frac{\Delta K}{\Delta t} = \dot{K} \qquad (2.4)$$

Where t is the time and a dot denotes change with respect to time. Substitution of (2.4) into (2.3) yields

$$S = \dot{K} \qquad (2.5)$$

Again, substituting (2.1) into (2.5), we obtain

$$sY = \dot{K}. \qquad (2.6)$$

In incremental or marginal terms, $v = \Delta K / \Delta Y$ or

$$v = \frac{\Delta K / \Delta t}{\Delta Y / \Delta t} = \frac{\dot{K}}{\dot{Y}} \qquad (2.7)$$

or

$$\dot{K} = v\dot{Y}. \qquad (2.8)$$

Substituting (2.8) into (2.6) we obtain

$$sY = v\dot{Y} \tag{2.9}$$

or

$$\frac{\dot{Y}}{Y} = \frac{s}{v}. \tag{2.10}$$

That is, the rate of growth of output is determined by the ratio between savings and capital–output ratios. It can also be shown that the rate of growth of capital stock is a constant and equal to s/v.

Replacing Y by K/v in (2.6) we obtain

$$\dot{K} = \left(\frac{s}{v}\right)K$$

Therefore,

$$\frac{\dot{K}}{K} = \frac{s}{v}. \tag{2.11}$$

Thus

$$\frac{\Delta Y}{Y} = \frac{s}{v} = \frac{\dot{K}}{K}. \tag{2.12}$$

Equation (2.12) is the 'fundamental' equation of the H–D model which indicates that, given the historically determined and constant values of s and v, the maximum rate of growth of capital stock is determined by the ratio s/v. This in turn determines the maximum attainable rate of growth of GDP *under the existing economic environment*.

In many developing countries, the savings rate s is quite low and v is high, implying inefficiency of investment. Hence the growth of the economy is low and may be insufficient to absorb a rapidly growing labour force.

It becomes necessary, therefore, to accelerate the growth of the economy beyond the limit set by the historical values of s and v. The planning problem begins here. It involves two issues: (i) fixing the target and (ii) determining how to achieve it.

Obviously the target growth rate must be sufficient to absorb the growing labour force. For example, if the population is expected to grow at 3 per cent per year and the planners would like to achieve a steady 4 per cent rise in per capita income then GDP must grow at 7 per cent annually. If we assume an aggregate capital–output ratio v

of 3 then for a 7 per cent target growth of GDP the required savings rate can be calculated from the fundamental equation (2.12) as

$$s^* = v\left(\frac{\dot{Y}}{Y}\right) = 3 \times 0.07 \times 0.07 = 0.21.$$

That is, 21 per cent of GDP must be saved so that actual saving is equal to the planned investment required to achieve a 7 per cent growth of GDP. This is the basis of Arthur Lewis's comment that the crux of the development problem is to raise the proportion of national income saved from 4–5 per cent to 12–15 per cent.

The next question, therefore, is how to raise the savings rate. There are three main sources of domestic savings: (i) personal savings, (ii) business savings and (iii) government savings. Most of the planning exercises emphasize the need to raise the government savings through taxation and thereby restrict consumption to fill the financial gap between actual savings and required investment. This presumes that savings cannot be raised significantly from the other two sources as they are largely voluntary in nature. But taxation is a financial matter which falls outside the domain of the planning agency. In most developing countries the planning and the budgetary management functions are directed by separate agencies. A common analytical framework is therefore needed to harmonize the concerns of the two agencies for the successful implementation of the plan, and to avoid any potential conflict of objectives.

The need for harmonization between the planning and the budgetary agencies becomes clearer if the planning problem is viewed from an alternative perspective. Since capital is seen to be the limiting factor to growth, the cost of capital (real interest rate) in the developing countries is often kept deliberately below the market rate. As Gurley and Shaw (1955) and McKinnon (1973) have pointed out, this deliberate credit rationing policy, instead of encouraging private investment, leads to the misallocation of investible funds and results in a higher capital–output ratio v. At the same time, a low (and in an inflationary situation, a negative) real interest rate discourages private savings, resulting in a low savings rate s. Thus, the Gurley–Shaw–McKinnon hypothesis provides us with a clue to the historically low s/v ratio. This policy-induced distortion is compounded in an underdeveloped economy characterized by the existence of a large non-monetized sector. In such an economy the absence of well-developed financial institutions means that a significant portion of the real investible surplus over consumption is not transformed into financial capital. Thus, two policy implications follow.

1 'Liberalize' the financial market, i.e. allow the interest rate to rise to its market equilibrium level. To the extent that the low value of s/v is due to policy-induced distortions, this will improve the economy's growth potential by encouraging private savings and improving the efficiency of capital.
2 Develop financial institutions which will mediate between savers and investors. Given adequate financial incentives, the accessibility to banks and other financial institutions is likely to tap the surplus over consumption which would otherwise have found its way into either hoarding or precious metals. Therefore, the programme for financial reforms should be an integral part of development planning.

If the mobilization of domestic savings through taxation and financial reforms is not enough to finance the required investment, the savings–investment gap can be closed by foreign savings. This can be seen quite clearly from the following familiar national accounts identity for an open economy:

$$Y = C + I + G + X - M \qquad (2.13)$$

$$Y = C + S + T \qquad (2.14)$$

where Y is gross national product (GNP), C is consumption expenditure, I is gross investment expenditure, G is government expenditure, X is the quantity of exports, M is the quantity of imports, S is the amount of savings, and T is the tax revenue. From (2.13) and (2.14) it follows that

$$I - [S + (T - G)] = M - X. \qquad (2.15)$$

That is,

investment–(private savings + government savings) =
current account deficit/surplus

From the above identity the financing options become clear. In a closed economy, if investment is to rise and private savings cannot be raised substantially then the only option left is to increase government savings through taxation. But in an open economy there is an additional option. The gap between the 'required' investment and 'actual' domestic savings can be financed by a deficit in the current account (the gap between imports and exports).

THE TWO-GAP MODEL

The previous section showed how foreign savings could meet a short-fall of domestic savings and enable the 'required' level of investment to be undertaken. Consideration of the role of foreign savings in the growth process is the main focus of the two-gap model developed by Chenery and associates (Chenery and Bruno 1962).

In most developing countries, not only does the level of domestic savings fall short of the required investment, but also export earnings are less than import expenditures. Foreign exchange is, therefore, required for the dual function of meeting both the 'trade gap' and the 'savings gap'. The contribution of the Chenery model is in identifying which of these gaps is the effective constraint on growth.

The basic two-gap model can be summarized as follows. We make the following assumptions.

1 There exists a fixed maximum propensity to save out of income, such that

$$\max S = sY \qquad 0 < s < 1 \qquad (2.16)$$

where s is the average and marginal propensity to save.
2 Certain capital goods can only be obtained from foreign sources.
3 There is a lack of substitutability between domestic and foreign resources. In particular, it is impossible to transform domestic savings into foreign savings via increased export flows. Given (2), this means that there is always a minimum amount of foreign exchange required to sustain growth.

Thus, the fixed proportion production function of the H–D model can be rewritten as

$$Y = \min\left(\frac{K_d}{k_1}, \frac{K_f}{k_2}\right) \qquad (2.17)$$

where $k_1 = K_d/Y$ is the amount of domestic capital required to produce one unit of output and $k_2 = K_f/Y$ is the amount of foreign capital required to produce one unit of output.

Given the lack of substitutability between domestic and foreign resources, the production function (2.17) implies that the growth will be constrained by either K_d (domestic capital) or K_f (foreign capital), whichever is the minimum. Unless the constraint is removed, a portion of the other factor (resource) will be unused.

To see the implications of the existence of a domestic savings or foreign exchange constraint, we can assume for simplicity that the country in question imports capital goods only, so that

$$M = I_f \tag{2.18}$$

where M is the quantity of imports and

$$I_f = \frac{\Delta K_f}{\Delta t} = \dot{K}_f. \tag{2.19}$$

Let us further assume that imports are a fixed proportion of GNP, such that

$$M = mY \tag{2.20}$$

From the supply side, export possibilities depend on the economy's output capacity, so that

$$\max X = xY \qquad 0 < x < 1. \tag{2.21}$$

Let us suppose that the growth rate permitted by foreign exchange is less than that permitted by domestic savings. Thus, from the production function (2.17), we have

$$Y = \frac{K_f}{k_2}.$$

Differentiating both sides of the above equation with respect to time t, we obtain

$$\frac{\Delta Y}{\Delta t} = \frac{1}{k_2} \frac{\Delta K_f}{\Delta t}$$

or,

$$\dot{Y} = \frac{1}{k_2} \dot{K}_f. \tag{2.22}$$

Substitution of (2.18), (2.19) and (2.20) into (2.22) yields

$$\dot{Y} = \frac{m}{k_2} Y$$

18

therefore,

$$\frac{\dot{Y}}{Y} = \frac{m}{k_2}. \tag{2.23}$$

That is, the growth of output is equal to the ratio between the import–output ratio m and the incremental foreign-investment-goods–output ratio k_2.

Similarly, if domestic savings is the limit to growth, we shall have

$$\frac{\dot{Y}}{Y} = \frac{s}{k_1} \tag{2.24}$$

which is exactly the same condition as the fundamental equation of the basic H–D model.

If the planners set a target growth rate of, say, g^*, the required savings ratio s^* and import ratio m^* can be calculated from equations (2.23) and (2.24) as:

$$s^* = g^* k_1 \tag{2.25}$$

and

$$m^* = g^* k_2. \tag{2.26}$$

If the maximum domestic savings is found to be less than the level required to achieve the target growth then there is said to exist an investment–savings gap equal to

$$I_d^* - \max S = S_d^* - \max S$$

(since in equilibrium I_d^* must equal S_d^*)

$$\begin{aligned} I_d^* - \max S &= s^* Y - s Y \\ &= k_1 g^* Y - s Y. \end{aligned} \tag{2.27}$$

Similarly, if the maximum export potential (earnings) is found to be less than the minimum import requirements for the target growth then there is said to exist an import–export gap equal to

$$M^* - \max X = k_2 g^* Y - x Y. \tag{2.28}$$

Even though, according to the national accounts identity, the *ex post* investment–savings gap must be equal to the *ex post* import–export gap, there is no reason for these two gaps to be equal *ex ante* (in a planned sense). Therefore, the investment–savings and import–export gaps (*ex ante*) can be written as

$$k_1 g^* Y - sY \leqslant F \qquad (2.27a)$$

and

$$k_2 g^* Y - xY \leqslant F \qquad (2.28a)$$

where F is the foreign capital (exchange) inflow.

Thus, it is intuitively clear that if the target growth is to be achieved, foreign capital inflow must fill the larger of the two gaps so that the bottleneck can be avoided. Developing countries are often found to suffer from the 'foreign exchange constraint' (i.e. the import–export gap is larger than the investment–savings gap). In this situation, the willingness to save is frustrated by the inability to acquire imports which are required to produce investment goods. That is, a portion of the maximum potential domestic savings remains unutilized in the presence of a constraint imposed by the availability of foreign exchange. This situation can be depicted graphically by plotting inequalities (2.27a) and (2.28a) in a g^*–F plane (Figure 2.1). To the left of F^*, the foreign exchange constraint is operative, a situation believed to be faced by a typical developing country. In such a situation, unless the constraint is removed, the growth will be lower

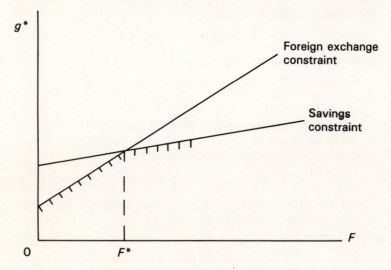

Figure 2.1 Two-gap model restrictions

20

than what the maximum potential domestic savings would have allowed. Herein lies, according to the two-gap model, the importance of foreign aid in the development process.

This conclusion, however, crucially depends on the assumption of no substitution between foreign and domestic resources; in other words, if foreign exchange is scarce it is impossible to use domestic resources to earn more foreign exchange. Critics of the two-gap model have argued that developing countries should be able to transform surplus domestic resources into export production. But, as Thirlwall (1983: 295) has put it: 'If it were that easy, the question might well be posed, why do most developing countries suffer from chronic balance of payments deficits over long periods despite vast reserves of unemployed resources?' Critics have also pointed out that, instead of supplementing domestic savings, foreign aid/exchange may substitute for it, if foreign aid relaxes the domestic savings effort. It is also argued that foreign aid has allowed governments to delay making difficult decisions regarding domestic resource mobilization.

FURTHER READING

Taylor (1979: ch. 8) gives a clear exposition of the Harrod–Domar and the two-gap models. Bhagwati and Chakravarty (1969) discuss the application of these planning models in India. Chenery and Bruno (1962) give an exposition of the two-gap models and their application to the Israeli economy.

3

DISAGGREGATED CONSISTENCY MODELS

INTRODUCTION

The planning exercise does not stop at the most aggregate level with the setting of a GNP growth target and the calculation of required investment, savings and foreign capital inflows. As mentioned earlier, the whole process proceeds in stages, and the second stage of the planning exercise usually involves a further breakdown of the plan to cover a number of strategic sectors of the economy. The economy is the sum of the individual sectors, so the target growth of GNP will imply specific rates of growth for the component sectors. The second stage therefore involves the translation of the overall growth target and investment requirements into sectoral growth targets and investment requirements. This is the concern of disaggregated consistency models. In this chapter we consider two such models: the main-sector model and the multi-sector model.

MAIN-SECTOR MODELS

Besides increasing the overall growth rate, development planning also aims at transforming the sectoral balance of the economy. Typically, this implies the transfer of resources from agriculture to industry as the economy grows. But at the same time, the agricultural productivity must improve so that industrial production is not hampered by shortages of raw materials, food and manpower. This means that at the second stage the planners must ensure that the target sectoral growth rates satisfy the conditions for a 'balanced' growth. The main-sector models enable the planners to break down the aggregate growth target and investment requirements into sectoral targets and requirements as well as to specify conditions for a 'balanced' growth.

Characteristics of the main-sector models

Even though the focus of the main-sector models is sectoral growth and investment, the scope still remains the overall economy. From this perspective the main-sector models must fulfil certain requirements. These are (i) completeness, (ii) consistency and (iii) economical and realistic requirements.

1. *Completeness*   The main-sector models are economy-wide models and hence when sectoring the economy no activity can be left unaccounted for. That is, all the activities must be allocated to a sector. For this reason, it is sometimes convenient to have a residual sector, 'the rest of the economy', which includes activities which cannot be allocated to the other main sectors.

2. *Consistency*   This requirement follows from (1). Since the coverage is the whole economy, the sectoral value added must add up to the GDP. In addition, the accounting relationships require that the total demand (final and intermediate) for a sector's output must not exceed its total output or supply.

3. *Economical and realistic requirements*   There are operational requirements. While the whole economy must be covered, the number of sectors should be kept within a manageable limit. As the term 'main sectors' indicates, the model should subdivide the economy into a few strategic sectors only and focus on essential relationships.

Examples of the main-sector models: consumption and investment goods sectors

In the first stage of the planning exercise with the aggregate models, we have seen the crucial role of capital accumulation and savings. Since savings is an abstinence from present consumption, it must be rewarded with a higher level of consumption in the future. That is, growth should ultimately mean an increased provision of consumption goods. Therefore, at the second stage of the planning exercise, the planners can divide the economy into an investment goods sector and a consumption goods sector. This will enable the planner to evaluate the trade-offs between present and future consumption.

If for simplicity we assume that there is no intermediate transaction between the two sectors and that they only purchase final goods from each other then the output of the two sectors will add up to the GDP. That is,

$$X_1 + X_2 = \text{GDP} \tag{3.1}$$

where X_1 is the output of the investment goods sector and X_2 is the output of the consumption goods sector.

Given the marginal propensity to consume (MPC), the supply of consumption goods should be sufficient to meet the demand for them. Thus,

$$X_2 = b(X_1 + X_2) \tag{3.2}$$

where b is the MPC. From equation (3.2) one can see that, for a given MPC, the output or productive capacity of the two sectors must bear a constant relationship to each other such that

$$\frac{X_2}{X_1} = \frac{b}{1 - b} \tag{3.3}$$

A typical planning problem, therefore, is to determine the distribution of investment goods between the two sectors that will ensure a 'balanced' growth of the sectors as determined by (3.3).

Given the target (desired) rate of growth of income, the prevailing consumption–income ratio (or MPC) determines the demand for consumption goods. But often this level tends to exceed the existing capacity of the consumption goods sector. Therefore, if shortages of consumption goods and resultant inflation are to be avoided, the production capacity of the consumption goods sector must be raised by new investments in that sector. But the new investment goods must be supplied by the investment goods sector and hence the investment goods sector must have the capacity to produce additional machines. Otherwise, the capacity of the machine-producing sector must be raised first. Obviously, in such a situation, the growth of present consumption must be restrained in order to provide funds for the initial expansion of the investment goods sector. This makes the distribution of investment a difficult choice.

This was the situation during the early days of Soviet planning and one of the most creative discussions of applied planning problems took place during the Soviet industrialization debate of the mid-1920s (for an excellent review of the debate, see Erlich 1960). The debate on the question of the distribution of capital goods when such goods are in short supply was formalized by the Soviet engineer–economist Feldman. This model was later modified by Domar (1957), and

Mahalanobis (1953) developed a very similar model where production of 'machines to produce machines' is the central issue. This model served as the intellectual underpinning in the early Five-Year Plans in India (Bhagwati and Chakravarty 1969).

The key assumption of the Feldman–Domar–Mahalanobis (F–D–M) model is that capital cannot be shifted away from the sector in which it is first installed. The main question it seeks to answer is: what fraction of investment goods should be assigned to the investment goods sector itself and how much to the production of new consumer goods? If k_1 is the capital–output ratio in the investment goods sector then, assuming full-capacity utilization, the output of capital goods will be

$$X_1 = \frac{K_1}{k_1} \qquad (3.4)$$

Let us assume that initially a fraction r of the newly produced capital goods is assigned to the investment goods sector for the expansion of the sector itself. Then the net addition to its capital stock (assuming no depreciation) will be

$$\dot{K}_1 = \frac{\Delta K_1}{\Delta t} = rX_1 = \frac{rK_1}{k_1}.$$

Therefore,

$$\hat{K}_1 = \frac{\dot{K}_1}{K_1} = \frac{r}{k_1} \qquad (3.5)$$

is the rate of growth of capital stock in the investment goods sector.

If rK_1/k_1 of capital goods is assigned to the investment goods sector then $(1 - r)K_1/k_1$ will be left to be assigned to the consumer goods sector. Therefore, the net addition to the consumer goods sector's capital stock (again, assuming no depreciation) is

$$\dot{K}_2 = \frac{\Delta K_2}{\Delta t} = (1 - r)\frac{K_1}{k_1}$$

$$\frac{\Delta K_2/\Delta t}{K_2} = \frac{1 - r}{k_1}\frac{K_1}{K_2} \qquad (3.6)$$

or

$$\hat{K}_2 = \frac{\dot{K}_2}{K_2} = \frac{(1 - r)}{k_1}\frac{K_1}{K_2}$$

is the rate of growth of capital stock in the consumer goods sector. Clearly, the growth of the consumption goods sector's capital stock depends on the composition of capital stocks (K_1/K_2) in the two sectors. If we plot the equations (3.5) and (3.6) we can see the implications of this relationship between the growth of capital stocks in the two sectors (Figure 3.1).

E_0 is the initial steady-state equilibrium where the growth rates of capital stocks in the two sectors are the same. This ensures that the output or capacity of the two sectors bears a constant relationship as indicated in equation (3.3).

If a more aggressive investment programme is adopted requiring the fraction of capital stock assigned to the investment sector to rise from r to r^* then the lines showing the growth of capital stocks will shift to new positions (broken lines). Thus, the growth rate of K_1 will be r^*/k_1. On the other hand, the line for K_2^* will be less steep, implying a fall in the growth rate of capital stock (and hence capacity) in the

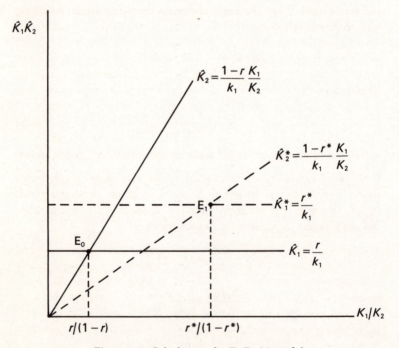

Figure 3.1 Solution to the F–D–M model
Source: Taylor (1979: 121)

consumption sector. However, ultimately the growth rate of capital stock in the consumption sector will rise to a new level, and a higher steady-state equilibrium will be established at E_1 when

$$\frac{K_1}{K_2} = \frac{r^*}{1 - r^*}. \tag{3.7}$$

For policy purposes, it is essential to ask what happens to consumption in the short run under the new investment plan. Obviously, during the period when the growth rate of the consumption goods sector falls, the growth of consumption must be restrained to keep it within the supply capacity of the sector. This essentially means a reduction in the consumption–income ratio or raising the savings rate to finance the aggressive investment programme. In terms of the 'balance' equation (3.3), it requires an alteration of the existing relationship between the two sectors (i.e. lowering the ratio $b/(1 - b)$) in favour of the investment goods sector. Therefore the savings ratio $(1 - b)$ becomes a policy parameter.

This brings out an important difference between the H–D model and the F–D–M model. While in the aggregate H–D model the savings rate is determined by behavioural characteristics of the decision-making agents, the F–D–M model treats it as a rigid function of certain structural features of the economy such as the capacity of the domestic investment and consumption goods sectors. This model therefore plays down the importance of savings as a limitation on growth; rather it regards the capacity of the investment goods sector as the constraining factor. By making the allocation ratio a policy variable, it shows that a higher allocation of investment goods to that sector itself would mean a higher savings rate at the margin and hence a higher rate of growth of output. In deriving this central proposition, the F–D–M model implicitly assumes that the government is in a position to control consumption completely. This is a weakness of the F–D–M model. Even if the government can control private consumption, the choice of too large a value of the allocation ratio and the consequent repression of consumption may be politically unacceptable. Moreover, the F–D–M model is a closed economy model and ignores an important escape route – foreign trade.

In fact, by relaxing the assumption that domestic and foreign resources are not substitutable, the two-gap model can be converted into a two-sector model with one sector for domestic consumption and the second for exports. As the experience of the East Asian newly industrializing countries shows, foreign revenues can be used to buy capital goods from abroad and the capacity of the domestic capital goods sector no longer remains a constraint. This extension brings out the importance of current investment and foreign borrowing to develop industries that will 'produce' dollars via exports for future use.

The central message of the main-sector models is that when there are shortages of some crucial factor (be it capital goods or foreign exchange), it is not efficient to invest in all the sectors. Rather the investment plan should concentrate on a few 'strategic' or 'key' sectors. A question, however, that arises is how to define or identify the 'key' sectors. One common practice is to define the 'key' sectors in terms of Hirschman's backward and forward linkages. The input–output technique, the topic of the next section, is widely used to identify the key sectors in this particular sense.

MULTI-SECTOR MODELS: INPUT–OUTPUT ANALYSIS

At the core of a multi-sector planning model is the input–output analysis. The essence of the input–output analysis is that it captures the interrelationships of the production structure arising through the flow of intermediate goods. There are numerous uses of the input–output technique and it is the most widely used planning tool. The most common uses of it for planning purposes can be summarized as follows.

1 Being a consistency model, the input–output technique can be used for projecting and forecasting sectoral production or supply requirements to meet the sectoral demands implied by alternative targets for GDP.
2 Once the sectoral supply is projected, the analysis can be extended to forecast the requirements for (a) primary factors, e.g. skilled labour, (b) sectoral capacity expansion and investment and (c) non-competitive imports.
3 Since the input–output table contains information on intermediate transactions between the producing sectors, it can be used to identify the 'key' sectors with maximum interdependence.
4 The input–output table can also be used to forecast the impact of supply shocks and cost inflation.
5 Finally, the input–output technique can be applied to determine comparative advantages.

Each of these uses will be discussed in greater detail at subsequent stages. But before we proceed to do so, it is germane to provide an outline of the input–output system.

Input–output: a sketch of the system

The input–output accounting framework can be represented in a schematic form, using three blocks (Table 3.1). Block I shows the supply of output from one sector to another for use in the production process of each sector. That is, this is the block of the table where sectoral outputs become inputs in other sectors. The rows show the distribution of output of a given sector and the columns show the structure of inputs for the same sector. Block II shows the final-demand use of the sectoral output, and block III records the primary factors and non-competitive imports.

Therefore, blocks I and II together give the distribution of sectoral output as intermediate and final demands. That is, if we denote the ith sector's intermediate delivery to the jth sector as X_{ij} and its delivery of the lth category of final demand as F_{il} then the sum of the ith row will be equal to the gross output X_i of the ith sector. This can be expressed algebraically as

$$\sum_j X_{ij} + \sum_l F_{il} = X_i. \tag{3.8}$$

The balance equation (3.8) ensures that the multi-sectoral input–output model satisfies the consistency requirements.

A central task in input–output work is the calculation of coefficients of production, known as input–output technical coefficients. These coefficients show inputs per unit of output for a given sector. Thus, we can derive the technical coefficients as:

$$a_{ij} = \frac{X_{ij}}{X_j}. \tag{3.9}$$

Table 3.1 A schematic input–output table

Sales	Purchases		
	Sales of i to j	*Industry final demand deliveries*	*Gross output*
	$[X_{ij}]$ Block I	$[F_{i1}]$ Block II	$[X_i]$
Value added	$[Y_{ik}]$ Block III		
Total inputs	$[X_j]$		

Substituting (3.9) into (3.8), the balance equation can be rewritten as

$$X_i = \sum_j a_{ij}X_j + \sum_l F_{il}$$

or

$$X_i = \sum_j a_{ij}X_j + F_i. \tag{3.10}$$

Assuming only two sectors (for simplicity), we can write the balance equation (3.10) more explicitly as

$$X_1 = a_{11}X_1 + a_{12}X_2 + F_1$$
$$X_2 = a_{21}X_1 + a_{22}X_2 + F_2. \tag{3.10a}$$

Collecting the common terms in the equations system (3.10a) and rearranging, we obtain

$$(1 - a_{11})X_1 - a_{12}X_2 = F_1$$
$$- a_{21}X_1 + (1 - a_{22})\, X_2 = F_2. \tag{3.10b}$$

In the input–output analysis, final demands F_i are assumed exogenous. Thus, given the values of final demands, the system (3.10b) can be solved for sectoral gross output levels X_i.

Solution of the system: the matrix method

The input–output system, or the equations system (3.10b), can be solved for sectoral gross output by using one of the following methods: (i) the iteration or successive approximation method, (ii) Cramer's rule or (iii) the matrix method.

For a larger system, methods (i) and (ii) can be quite cumbersome. The usual practice is therefore to solve the system by using the matrix method. (The essentials of matrix algebra are presented in the appendix to this chapter.) Besides its computational advantages, the matrix method yields some useful corollaries. For example, with the matrix method it is easier to analyse the chain of 'direct' and 'indirect' effects of an initial shock or change.

The equations system (3.10b) can be rewritten in matrix form as

$$\begin{bmatrix} 1 - a_{11} & - a_{12} \\ - a_{21} & 1 - a_{22} \end{bmatrix} \begin{bmatrix} X_1 \\ X_2 \end{bmatrix} = \begin{bmatrix} F_1 \\ F_2 \end{bmatrix}.$$

Therefore,

$$(I - A)X = F \qquad (3.10c)$$

where A is the matrix of technical coefficients a_{ij} and is known as the 'direct' requirements matrix.

From simple algebra, we know that if

$$ax = y$$

then

$$x = y/a = a^{-1}y.$$

Using a similar but not exactly equivalent method, we can find from (3.10c) the sectoral gross output levels X for given values of final demands F as

$$X = (I - A)^{-1}F \qquad (3.10d)$$

where $(I - A)^{-1}$ is known as the 'Leontief inverse' or 'direct and indirect' requirements matrix.

Written more explicitly, the solution for X_1 is

$$X_1 = [(1 - a_{11})(1 - a_{22}) - a_{12}a_{21}]^{-1} \, [(1 - a_{22})F_1 + a_{12}F_2].$$

The term in the first bracket is the determinant $|I - A|$. It measures the ability of the economic system to produce a surplus for final demand in excess of current intermediate uses. For an economically viable system, $|I - A|$ should be positive. This is satisfied if each column sum in A is less than unity. This is known as the 'stability condition'. By stability we mean that if the system is disturbed from an equilibrium by 'shocks' (in this case, an exogenous change in final demand), it will again reach an equilibrium.

From the explicit solution for X_1 above, we can see that X_1 exceeds F by a multiple of the Leontief inverse. From (3.10d) we can obtain

$$\frac{\Delta X}{\Delta F} = (I - A)^{-1}. \qquad (3.11)$$

Therefore, the Leontief inverse gives the direct and indirect impacts of an exogenous change in final demand. Hence, it can be regarded as the 'input–output multiplier'. If we denote the elements of $(I - A)^{-1}$ by r_{ij} then (3.11) can be rewritten explicitly as

$$\Delta X_1 = r_{11} \, \Delta F_1 + r_{12} \, \Delta F_2 \qquad (3.11a)$$

$$\Delta X_2 = r_{21} \, \Delta F_1 + r_{22} \, \Delta F_2. \qquad (3.11b)$$

Assuming an exogenous change in the final demand for sector 1's output only (i.e. $\Delta F_2 = 0$), we can obtain (from (3.11a) and (3.11b)) the total changes in all the sectoral output as

$$\Delta X_1 + \Delta X_2 = (r_{11} + r_{21})\Delta F_1$$

such that

$$\frac{\Delta X_1 + \Delta X_2}{\Delta F_1} = r_{11} + r_{21}.$$

Thus, the first column sum of the Leontief inverse gives the change in total output from all sectors to meet the direct and indirect requirements arising from a unit increase/decrease in final demand for the first sector's output. In general, therefore, the ith column sum of $(I - A)^{-1}$ gives the multiplier effect of a unit change in the ith sector's final demand delivery. On the other hand, the jth row sum of $(I - A)^{-1}$ gives the multiplier effect of a unit change on the jth sector only when final demands for all sectoral outputs change simultaneously by one unit each.

A Numerical Example

In Table 3.2 we present hypothetical inter-industry transactions for only two sectors, say agriculture and manufacturing. It should be mentioned that these two sectors must cover all production activities in the entire economy.

Table 3.2 A hypothetical input–output transactions matrix

		Purchases		
Sales	Agriculture	Manufacturing	Final demand	Total output
Agriculture	10	50	40	100
Manufacturing	80	60	60	200
Value added	10	90		
Total input	100	200		

The coefficients matrix is derived as

	Agriculture	Manufacturing
Agriculture	10/100 = 0.10	50/200 = 0.25
Manufacturing	80/100 = 0.80	60/200 = 0.30

Therefore,

$$A = \begin{vmatrix} 0.10 & 0.25 \\ 0.80 & 0.30 \end{vmatrix}$$

and

$$(I - A)^{-1} = R = \begin{vmatrix} 1.627 & 0.58 \\ 1.86 & 2.1 \end{vmatrix}.$$

Using the general solution (3.10d), we obtain

$$X_1 = 1.627 \times 40 + 0.58 \times 60 = 101.6$$

and

$$X_2 = 1.86 \times 40 + 2.1 \times 60 = 200.4.$$

Now if the demand changes, the output effects can be calculated by again solving (3.10d) for the new levels of final demand. Alternatively, one can use the multiplier relationship (3.11) to determine the change in output levels. For example, if final demand for agriculture increases by, say, 40 units, then the new solutions will be

$$X_1^* = 1.627 \times 80 + 0.58 \times 60 = 165$$

and

$$X_2^* = 1.86 \times 80 + 2.1 \times 60 = 275.$$

Therefore,

$$\Delta X_1 = X_1^* - X_1 = 63.4$$

$$\Delta X_2 = X_2^* - X_2 = 76.6.$$

Alternatively, by using (3.11) we can obtain

$$\Delta X_1 = 1.627 \times 40 = 63.4$$

$$\Delta X_2 = 1.86 \times 40 = 76.6.$$

It is interesting to note that the main diagonal elements of $(I - A)^{-1}$ are greater than unity. This arises from the economic system's ability to produce a surplus for final demand in excess of current intermediate uses. That is, when final demand for a particular sector increases by one unit, the sector must produce that one unit and whatever is 'directly and indirectly' needed for intermediate use to produce the required additional unit. Thus, in our example,

$r_{11} = 1.627$ implies that sector 1 must produce an excess of 0.627 units which is 'directly and indirectly' required for intermediate use to produce an additional unit in sector 1 to satisfy final demand. In other words, in the process of producing one unit of final output, it generates 0.627 units of demand for intermediate inputs from itself to be used by all sectors (including itself). Thus, in 1.627, 1 is the output effect and 0.627 is the 'direct and indirect' input effect.

To illustrate this process of output and input effects we shall use the iterative method. An increase in final demand of 40 from the agricultural sector requires additional direct inputs from itself and the manufacturing sector by amounts equal to the technical coefficients multiplied by the additional output required. That is, the 'direct' input effects are

$$0.10 \times 40 = 4 \qquad \text{extra units from agriculture}$$

$$0.80 \times 40 = 32 \qquad \text{extra units from manufacturing.}$$

In order to produce these additional outputs for intermediate use, further output increases due to indirect input effects will be required in the following manner.

Round 1:

	Agriculture	Manufacturing	Total
Agriculture	$0.10 \times 4 = 0.4$	$0.25 \times 32 = 8.0$	8.4
Manufacturing	$0.80 \times 4 = 3.2$	$0.30 \times 32 = 9.6$	12.8

Round 2:

	Agriculture	Manufacturing	Total
Agriculture	$0.10 \times 8.4 = 0.84$	$0.25 \times 12.8 = 3.2$	4.04
Manufacturing	$0.80 \times 8.4 = 6.92$	$0.30 \times 12.8 = 3.84$	10.76

and so on.

One can observe that at the second round the total increment in each sector has declined. In fact, the total increment at every successive stage will decline. This convergence occurs because the technical coefficients are less than unity. Thus, we can derive the total change in agricultural output as

$$X_1 = 40 + 4 + 19 \approx 63.4$$

$$= \text{Change in final demand} + \text{Direct input effect} + \text{Indirect input effects}$$

and for the manufacturing sector

$$X_2 = 0 + 32 + 43 \approx 76.6$$
$$= \text{Direct input effect} + \text{Indirect input effects}$$

POLICY APPLICATIONS OF INPUT–OUTPUT ANALYSIS

Forecasting sectoral output

We saw earlier that for a given level of final demand there exists a unique set of sectoral gross outputs. In the context of planning, the levels of various categories of final demand are determined at the first stage of the exercise. Given the desired growth rate of GDP and its terminal value, the planners can work out the various categories of sectoral final demand by using some form of Engel curve relating expenditure to income. For example, the terminal value of the ith sector's delivery to consumption can be determined in two steps. The first step involves the estimation of an aggregate consumption function $C = f(Y)$ (where C is aggregate consumption and Y is GDP) using time series data from the national accounts. The estimated parameters (e.g. MPC) of the consumption function are used to obtain the terminal value of aggregate consumption for the terminal value of GDP. The next step is to allocate the estimated terminal value of consumption to sectoral final demand deliveries. This can be done by assuming that the base year distribution of sectoral final demand deliveries remains constant over the plan period. For a medium-term plan, this does not appear to be too unrealistic an assumption. Therefore, we can define a constant

$$h_i = \frac{C_i}{C} \tag{3.12}$$

such that $\Sigma_i h = 1$ and where C_i is the base year delivery of consumption goods from the ith sector.

Thus, we can obtain sectoral deliveries of consumption goods in the terminal year as

$$C_i^* = h_i C^* \tag{3.13}$$

This exercise is carried out for other categories of final demand and we can rewrite (3.13) in general form as

$$F^* = HG^* \tag{3.14}$$

35

where F^* is an $n \times 1$ column vector of sectoral final demand deliveries in the terminal year, G^* is an $m \times 1$ column vector of aggregate final demand in the terminal year and H is an $n \times m$ base year sectoral distribution of final demand matrix.

The terminal value of sectoral final demand deliveries, thus obtained, can be used in the input–output relation (3.10d) to find the required level of sectoral gross output in the terminal year. Alternatively, by using the relationship (3.11), we can determine the required increase in sectoral output to sustain the increase in final demand in the terminal year.

This exercise, known as consistency or multiplier analysis, is usually carried out for a set of final demands derived from alternative estimates of desired GDP growth rate. This is necessary because each set of alternative estimates would have different implications for resource use and given the availability of resources some targets may not be feasible. Therefore, the next job of the planners is to estimate the resource requirements.

Forecasting primary factor or resource requirements

The information recorded in the third block of the input–output table (Table 3.1) is used for this purpose. From the third block we know how much is paid to different factors. Therefore, by dividing the factor payments by the given factor price, one can easily obtain the amount of a particular factor used to produce a given level of sectoral output. For example, if V_i is the payment to a factor, say skilled labour S_i, by the ith sector, then the total employment of skilled labour in the ith sector will be

$$S_i = \frac{V_i}{w} \tag{3.15}$$

where w is the wage rate for the skilled labour. Therefore the employment of skilled labour per unit of ith sectoral output will be

$$s_i = \frac{S_i}{X_i}. \tag{3.16}$$

Assuming that the skilled labour coefficients for different sectors remain constant over the medium-term plan period, one can determine the sectoral skill requirements for the terminal year sectoral output as

$$S_i^* = s_i X_i^* \tag{3.17}$$

$$S^* = sRF^* \tag{3.17a}$$

where $R = (I - A)^{-1}$ and s is an $n \times n$ matrix with s_i in the main diagonal and zeros elsewhere. Thus, (3.17a) gives estimates of skilled labour requirements in each sector to sustain the terminal year final demand F^* and the consequent sectoral gross output X^*.

This estimate is then compared with the projected supply of skilled manpower to see whether the overall output as well as sectoral output targets are feasible. Given the existing supply and anticipated increase in skilled manpower, if the target is not found to be feasible then the planners will have two options: (a) revise the target downward or (b) undertake a vigorous and comprehensive manpower development programme. In fact, this is the stage where 'special planning models' are applied for specific requirements. The input–output-based manpower or human resource development model is known as the manpower requirement approach.

Forecasting capacity expansion and sectoral investment requirements

Besides primary factors, the expansion of sectoral output to meet the terminal year final demand may require expansions of sectoral capacity. As we have seen in the main-sector models, additional capital goods may have to be installed in at least certain sectors. To determine the additional capital requirements we need information on sectoral incremental capital–output ratios (ICORs). This can be obtained from the information on inter-sectoral capital goods transactions. If, for example, K_{ij} is the amount of capital goods going from the ith sector to the jth sector then the sectoral capital coefficients will be

$$k_{ij} = \frac{K_{ij}}{X_j}.$$

Therefore, the ICOR for the jth sector is

$$\text{ICOR}_j = \sum_i k_{ij} = \frac{\Delta \sum_i K_{ij}}{\Delta X_j}.$$

From the sectoral ICOR we can obtain the total change in capital requirements for the jth sector as

$$\Delta K_j = \Delta \sum_i K_{ij} = \text{ICOR}_j \, \Delta X_j. \tag{3.18}$$

Since $\Delta X^* = (I - A)^{-1} \Delta F^*$, we can rewrite (3.18) in terms of the projected increase in final demand as

$$\Delta K = QR \, \Delta F^* \tag{3.18a}$$

where $R = (I - A)^{-1}$ and Q is an $n \times n$ matrix with sectoral ICORs in the main diagonal and zeros elsewhere.

Therefore, assuming the constancy of the ICOR over the relevant time frame, the planners can estimate the sectoral needs for additional capital goods or investment requirements for the projected increase in final demand by using the relationship (3.18a). This estimation forms the basis for detailed sectoral investment programmes. If the total requirement for capital goods is beyond the existing capacity of the capital-goods-producing sectors, then, of course, initially the investment programme will have to concentrate on the capital-goods-producing sectors themselves. In such a situation, the planners can make use of F–D–M type models.

Forecasting non-competitive import requirements and analysing possibilities of import substitution

Since foreign exchange can be an effective constraint for economic growth, economic development programmes have often been concerned with saving foreign exchange through import substitution industrialization (ISI). But it has frequently been alleged that ISI leads to a greater import dependence, the very phenomenon it wants to reduce. This is because, in the first phase, ISI substitutes imports of final products with domestic production. As the final demand shifts from the external source to the internal source, domestic output needs to be raised. This, in turn, generates direct and indirect demands for imported raw materials and capital goods. As a result, savings of foreign exchange from the substitution of the final products may be outweighed by the loss of foreign exchange through imports of raw materials and capital goods. The input–output analysis is a useful tool to examine the net foreign exchange savings of ISI. This can be demonstrated as follows.

If M_i is the amount of imports used in the ith sector as its inputs then we can define the sectoral import coefficients as

$$m_i = \frac{M_i}{X_i}. \tag{3.19}$$

Since $\Delta X^* = (I - A)^{-1} \Delta F^*$, the equations system (3.19) can be rewritten as

$$\Delta M^* = m(I - A)^{-1}\Delta F^*$$
$$= mR \, \Delta F^* \tag{3.19a}$$

where m is an $n \times n$ matrix with sectoral import coefficients as diagonal elements and zeros elsewhere.

Assuming constant import coefficients, (3.19a) gives the sectoral import requirements arising from a projected increase in final demand. This therefore provides an estimate of the balance-of-payments consequence of the development programme.

Now, if because of ISI one unit of final demand switches from imports to the jth sector, then the net savings will be

$$1 - \sum_i z_{ij} \tag{3.20}$$

where the z_{ij} are elements of $m(I - A)^{-1}$ such that $\Sigma \, z_{ij}$ is the direct and indirect import requirement in all the sectors arising from an increased delivery of final demand from the jth sector to replace imports.

Therefore, for a given exchange rate, (3.20) gives the net foreign exchange savings from import substitutions. If the net foreign exchange savings is positive for a particular sector then it merits import substitution. In fact, all the prospective sectors can be ranked in terms of net foreign exchange savings for the purpose of 'selective' ISI.

The exercises given in the previous four sections are primarily aimed at checking the consistency and feasibility of the overall growth objective. Given the existing sectoral capacity and resource availability, the initial target of GDP growth and the consequent estimates of final demand and sectoral gross outputs may have to be revised. This therefore shows how the different stages of planning are interrelated and how the feedback among them can improve the overall plan estimates.

Identification of 'key' sectors

Earlier we have seen the key role of the capital goods and export sectors in the main-sector models. It was mentioned that these models can provide the basis for 'unbalanced' growth. The proponents of the 'unbalanced' growth doctrine (e.g. Hirschman 1958; Rostow 1960) maintained that since in less developed countries (LDCs) certain critical factors, e.g. capital, are in short supply, it is not possible for them to start the process of development with a 'big push' by investing in all the sectors of the economy. They therefore suggested that the development programme should initially concentrate on certain 'key' sectors which due to their high degree of intersectoral dependence produced a 'carry-over' of growth impulses. Thus, a 'key' sector can be defined as a sector whose growth will promote or generate growth in other sectors via its technological ties with them. Growth impulses generated by a key sector encourage the expansion of other sectors in two main ways: (i) as a user of inputs from other sectors (backward linkage) and (ii) as a supplier of inputs to other sectors (forward linkage).

An expansion of a key sector having an above average use of inputs from other sectors will induce others to expand to meet its input requirements. This in turn will generate a further impact on those sectors which supply inputs to the sectors linked to the key sector and so on. Depending on the strength of interdependence, there could be a long chain of indirect impacts.

On the other hand, an expansion of a key sector which is an above average supplier of inputs is likely to lower the input price. This, in turn, is expected to encourage its users to expand. More importantly, in an economy plagued with shortages of all kinds, an increase in mere availability of inputs is itself an encouraging factor.

Since the 'Leontief inverse' $(I-A)^{-1}$ captures the direct and indirect input requirements, it can be used to identify key sectors with above average linkage effects. As we know, the ith column sum of $(I-A)^{-1}$ gives the total (direct plus indirect) input requirements from all the sectors for a unit increase in final demand for the ith sector's output. And the ith row sum of $(I-A)^{-1}$ indicates the total (direct plus indirect) increase in the ith sector's output to meet a unit increase in final demand for all sectoral output. Therefore, averages of these direct and indirect impacts can be used as indicators of the degree of interdependence of a given sector. By relating these averages for individual sectors to the overall (economy-wide) average, we can rank the sectors in terms of their degree of interdependence. Thus, by concentrating on relationships between column elements of $(I - A)^{-1}$ we obtain

$$\text{backward linkage} = \frac{\Sigma_i \, r_{ij}/n}{\Sigma_i \, \Sigma_j \, r_{ij}/n^2}$$

and by concentrating on relationships between row elements we obtain

$$\text{forward linkage} = \frac{\Sigma_j \, r_{ij}/n}{\Sigma_i \, \Sigma_j \, r_{ij}/n^2}$$

where the r_{ij} are elements of $(I - A)^{-1}$ and n is the number of sectors.

Since the above indices are based on a method of averaging, a sector can have a high ranking in terms of either backward or forward linkage or both, but only be related to a limited number of sectors. Therefore, these indices are supplemented with the coefficient of variation (standard deviation/mean) to indicate how evenly the impulse is spread. The coefficient of variation for the backward linkage is

$$V_j = \frac{\{[\, 1/(n-1)]\Sigma_i \, (r_{ij} - \Sigma_i \, r_{ij}/n)^2\}^{1/2}}{\Sigma_j \, r_{ij}/n}.$$

The coefficient of variation for the forward linkage is

$$V_i = \frac{\{[\, 1/(n-1)]\Sigma_j \, (r_{ij} - \Sigma_j \, r_{ij}/n)^2\}^{1/2}}{\Sigma_j \, r_{ij}/n}.$$

A 'key' sector is therefore defined as one which has backward and forward linkages greater than unity and a relatively low coefficient of variation.

Analysing supply shocks and cost inflation

The input–output technique has been found to be extremely useful in analysing the impact of supply shocks. In the context of a feasibility check, we have seen earlier how the availability of resources (primary factors and non-competitive imports) can affect plan estimates. Unforeseen short-falls in the availability of resources can have pervasive effects on the price level by raising the cost of production. Therefore, together with the feasibility study, the planners should also carry out a sensitivity analysis of probable short-falls of critical supplies and their likely impact on costs of production. To illustrate the use of input–output analysis for tracking the impact of supply shocks, we shall assume (for simplicity) that labour is the only primary factor.

As we know, by reading down a column of an input–output table we obtain the corresponding sector's input structure. Thus, the jth column of the coefficient matrix would give the following decomposition of costs per unit of output in sector i:

$$P_j = \sum_i a_{ij}P_i + l_jW + m_jP_m \qquad (3.21)$$

where W is the wage rate, P_m is the price of non-competitive imports, l_j is the labour coefficient L_j/X_i and m_j is the import coefficient M_jX_i. That is,

cost-determined output price = per unit raw material costs + per unit primary factor costs + per unit import costs.

The cost-determined sectoral price equations system (3.21) can be rewritten as

$$P_1 = a_{11}P_1 + a_{21}P_2 + l_1W + m_1P_m$$
$$P_2 = a_{12}P_1 + a_{22}P_2 + l_2W + m_2P_m. \qquad (3.21a)$$

Given the prices of primary factors and non-competitive imports the system (3.21a) can be solved for sectoral cost-determined (or supply) prices. By using matrix notation we can rewrite (3.21a) as

$$P^T(I - A) = Wl^T + P_m m^T \qquad (3.22)$$

where the superscript T denotes the transpose of a matrix. Therefore,

$$P^T = (Wl^T + P_m m^T)(I - A)^{-1} \qquad (3.23)$$

gives the sectoral prices for the given prices of primary factors and non-competitive imports.

Equation (3.23) can be rearranged as

$$P^T = Wl^T(I - A)^{-1} + P_m m^T(I - A)^{-1}. \qquad (3.24)$$

Since $(I - A)^{-1}$ gives the direct and indirect material requirements per unit of final output, each entry in the row vector $Wl^T(I - A)^{-1}$ is the direct and indirect labour cost per unit of output in the corresponding sector. A similar statement applies to the import term $P_m m^T(I - A)^{-1}$. Thus, using relations like (3.24), the planners can find out the likely inflationary impact of probable supply shocks. For example, $P_m m^T (I - A)^{-1}$ captures the total (direct and indirect) inflationary impact of import cost increases on sectoral prices.

Identification of sectors with comparative advantage

The linkage approach to 'key' sectors probably applies well when a country follows an inward-looking strategy. But for an open or export-oriented economy, trade becomes the engine of growth. Hence, industrialization should be based on the principle of comparative advantage. A sector will have comparative advantage if the 'domestic resource cost' (DRC) of producing its product is less than the net foreign exchange that it can earn.

The input–output technique can also be used to calculate the sectoral DRCs. To illustrate, we will assume (for simplicity) that there is no protection (or tariff) on raw material imports. Therefore, the per unit domestic costs of raw material imports will be

$$P_m = eP_m^w \qquad (3.25)$$

where e is the exchange rate (price of foreign currency in local currency) and P_m^w is the world price of imports. Similarly, the per unit foreign exchange earnings from the ith sector will be

$$eP_i^w$$

where P_i^w is the world price value of product i.

Substituting the value of P_m from (3.25) into (3.23), we obtain the DRC for the ith sector as

$$DRC_i = (Wl^T + eP_m^w m^T)(I - A)^{-1} d_i \qquad (3.26)$$

where d is the column vector with ith element equal to one and all other elements zero.

A sector will have comparative advantage if

$$eP_i^w > (Wl^T + eP_m^w m^T)(I - A)^{-1} d_i . \qquad (3.27)$$

Equation (3.27) can be rearranged as

$$e > \frac{Wl^T (I - A)^{-1} d_i}{P_i^w - P_m^w m^T (I - A)^{-1} d_i} . \qquad (3.28)$$

The numerator of (3.28) is the purely domestic component of direct and indirect costs of producing the ith product. The denominator, on the other hand, is the 'net' foreign exchange gain from its exports. Therefore, if the direct and indirect domestic unit cost of producing the ith product relative to its net foreign exchange gain does not exceed the exchange rate then it is profitable to have the ith sector. Furthermore, if the factor prices are 'shadow prices' in the sense of being derived from some kind of social optimality criterion then the DRC is claimed to be the correct general equilibrium project selection criterion in an open economy. Methodologies of project analysis and the concept of shadow prices are discussed in subsequent chapters.

EXTENDING THE MODEL

The discussion in the preceding sections showed how the open static input–output method can be used for analysing a variety of planning problems related to economic growth. However, the input–output data system is lacking in an important aspect of development policy, i.e. the distribution of the benefits of growth. Final demands are treated as exogenous in the open input–output system (often because of a lack of information regarding the incomes of the individuals in the underlying data system), and the input–output scheme does not link functional or factoral income distribution to the household or institutional distribution. As a result, it misses the interrelationship between production (supply), distribution and expenditure (demand). However, it is the institutional income distribution with which plans are often concerned.

The flow diagram in Figure 3.2 presents the causal relationships between production, distribution and expenditure. Starting at any point in the triangle, the consistency of the system will ensure that the

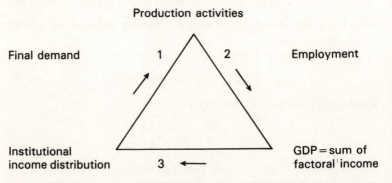

Figure 3.2 The causal relationships between production, distribution and expenditure
Source: Pyatt and Thorbecke (1976: 19)

feedback mechanism between the blocks yields the same initial set of values. For example, let us begin with a given institutional income distribution. This will generate a particular demand structure which, in turn, will determine the production activities (via link 1). The level of production activities thus determined will translate (via link 2) into derived demand for factors and factoral income distribution. Factor incomes then translate (via link 3) into institutional income on the basis of ownership. The consistency of the system requires that the derived institutional income distribution must be the same as the distribution assumed initially. That is, the economic system has the tendency to regenerate the initial income distribution.

In the conventional input–output data system, link 3 is not included. This raises two interrelated problems. First, by ignoring the mapping of GDP into individual income distribution, it is conceptually preoccupied with growth. But under conditions of initial uneven distribution of physical and human capital, the growth of GDP can ultimately result in a deterioration of the distribution of wealth and income. Since the consumption basket of the poor is different from that of the rich, the neglect of the income distribution link at the micro level will reproduce a production structure which will become increasingly biased against the basic needs of a large part of the population. Second, at the macro level, since the marginal propensity to consume of the majority of the people (who are poor) is larger (almost equal to unity) than the minority rich, the neglect of income distribution may lead to the problem of underconsumption or lack of absorptive capacity. Analytically, by treating final demands (expenditure) as exogenous, the input–output model short-circuits the traditional multiplier process (in the Keynesian sense) and fails to answer whether sufficient income is generated to the mass of people to pay for the products being produced.

FURTHER READING

Manne (1974) provides an excellent survey of multi-sector planning models, and Robinson (1989) gives a useful summary of the uses of multi-sector models. Blitzer *et al.* (1975: ch. III and V) give a comprehensive discussion of input–output models. Bulmer-Thomas (1982) provides a comprehensive treatment of the construction and use of input–output tables in developing countries. Dervis *et al.* (1982) is a more advanced text which applies the input–output methodology to the study of sources of structural change in a group of developing economies.

APPENDIX: ESSENTIALS OF MATRIX ALGEBRA

Matrix algebra is a useful tool for finding values for variables which are related by a system of linear equations. This appendix draws on Weber (1982).

Definitions

Matrix

A matrix is a rectangular array of numbers written in the form

$$A = \begin{bmatrix} a_{11} & a_{12} & \cdots & a_{1n} \\ a_{21} & a_{22} & \cdots & a_{2n} \\ \vdots & \vdots & & \vdots \\ a_{m1} & a_{m2} & \cdots & a_{mn} \end{bmatrix}.$$

The a_{ij} are real numbers, called the elements of the matrix, and i and j define their position in terms of row and column respectively. For example, a_{12} is the element in the first row and second column. In shorthand, the matrix A is denoted by (a_{ij}) and the numbers of rows and columns define the dimension or order of the matrix. For example, a matrix with m rows and n columns is referred to as an $m \times n$ dimensional matrix. If $m = n$, the matrix is said to be square. Two matrices of the same order are said to be equal if and only if all the corresponding elements are equal. For example, $A = B$ if and only if

$$A = \begin{bmatrix} 5 & 1 & 2 \\ 4 & 7 & 3 \end{bmatrix} \qquad B = \begin{bmatrix} 5 & 1 & 2 \\ 4 & 7 & 3 \end{bmatrix}.$$

Vector

A vector is a set of numbers arranged in either a column or a row. If they are arranged in a column it is called a column vector, and if arranged in a row it is a row vector. Thus,

$$U = \begin{bmatrix} u_1 \\ u_2 \\ \vdots \\ u_m \end{bmatrix}$$

is an m-dimensional column vector and

$$V = [v_1 \ v_2 \ldots v_n]$$

is an n-dimensional row vector. Therefore, a matrix can be regarded as being composed of a series of either column or row vectors.

Scalar

A real number which is a 1 × 1 matrix is called a scalar when it occurs in operations involving matrices. Thus, any constant that is used either to multiply (divide) or add to (subtract from) the elements of a matrix is a scalar.

Matrix operations

Operations such as additions, subtractions and multiplications can be performed with two or more matrices. However, since a matrix is an array of numbers, rather than a single number, some of the properties of operations involving single numbers do not hold for the analogous matrix operations. In the first place, matrices must conform to certain rules to be amenable for such operations.

Addition and subtraction

Two matrices can be added (subtracted) if and only if they are of the same order or dimension. The sum or difference of two matrices of the same order (say $m \times n$) is the sums or differences of the corresponding elements and will have the same order ($m \times n$) as the original matrices. Thus,

$$A \times B = \begin{bmatrix} a_{11} \pm b_{11} & a_{12} \pm b_{12} \\ a_{21} \pm b_{21} & a_{22} \pm b_{22} \end{bmatrix}$$

where

$$A = \begin{bmatrix} a_{11} & a_{12} \\ a_{21} & a_{22} \end{bmatrix} \qquad B = \begin{bmatrix} b_{11} & b_{12} \\ b_{21} & b_{22} \end{bmatrix}.$$

Multiplication by a Scalar

When a matrix is multiplied by a scalar, every element in the matrix is multiplied by that scalar (constant). Thus,

$$kA = \begin{bmatrix} ka_{11} & ka_{12} \\ ka_{21} & ka_{22} \end{bmatrix}$$

where k is a scalar (constant).

Multiplication of matrices

Two matrices A and B are said to be conformable for multiplication if and only if the number of columns in A is equal to the number of rows in B. Because of this conformability condition, the sequence in which matrices are multiplied is important. For example, if A is an $m \times n$ matrix and B is an $n \times p$ matrix then we can obtain AB but not BA. In the matrix product AB, A is said to premultiply B and B to postmultiply A. The product matrix AB will have the same number of rows as A and the same number of columns as B. It follows that when a $1 \times n$ row vector is premultiplied by an $n \times 1$ column vector, the result is a scalar (1×1 matrix). Thus, a scalar is the inner product of two vectors, i.e. the sum of products of the components of the vectors. Thus, if

$$U = [u_1 \quad u_2] \qquad \text{and} \qquad V = \begin{bmatrix} v_1 \\ v_2 \end{bmatrix}$$

then $UV = [u_1v_1 + u_2v_2] = w$ (a scalar).

Since a matrix is composed of a series of row and column vectors, the product matrix AB is obtained from the inner product of row vectors in A and column vectors in B. That is, the element in the ith row and jth column of AB is obtained from the inner product of the ith row in A and the jth column in B. For example, if

$$A = \begin{bmatrix} a_{11} & a_{12} \\ a_{21} & a_{22} \end{bmatrix} \qquad B = \begin{bmatrix} b_{11} & b_{12} \\ b_{21} & b_{22} \end{bmatrix}$$

then

$$AB = \begin{bmatrix} a_{11}b_{11} + a_{12}b_{21} & a_{11}b_{12} + a_{12}b_{22} \\ a_{21}b_{11} + a_{22}b_{21} & a_{21}b_{12} + a_{22}b_{22} \end{bmatrix}.$$

Even when both AB and BA are defined, in general $AB \neq BA$. For example, if

$$A = \begin{bmatrix} 5 & -6 \\ 1 & 0 \\ 0 & 3 \end{bmatrix} \qquad B = \begin{bmatrix} -1 & 8 & -3 \\ 0 & 10 & -4 \end{bmatrix}$$

then

$$AB = \begin{bmatrix} -5 & -20 & 9 \\ 1 & -8 & 3 \\ 0 & 30 & -12 \end{bmatrix} \qquad BA = \begin{bmatrix} -13 & -3 \\ -10 & -12 \end{bmatrix}$$

Therefore, $AB \neq BA$.

Although the sequence in which two matrices are multiplied affects the result, like the multiplication of single numbers, the order in which three or more matrices are multiplied does not affect the result, provided the sequence is maintained. That is,

$$A_{m \times n}B_{n \times p}C_{p \times c} = A_{m \times n}(B_{n \times p}C_{p \times q})$$
$$= (A_{m \times n}B_{n \times p})C_{p \times q}.$$

In summary, addition of matrices is cumulative, i.e. $A + B = B + A$, and both addition and subtraction are associative, i.e. $A \pm B \pm C = A \pm (B \pm C) = (A \pm B) \pm C$. Multiplication of matrices is not cumulative, i.e. $AB \neq BA$, but is associative, i.e. $ABC = A(BC) = (AB)C$. However, for single numbers, addition, subtraction and multiplication are both associative and cumulative.

The transpose of a matrix

The transpose of a matrix is obtained by changing its rows into columns and columns into rows. Thus if

$$A = \begin{bmatrix} 3 & -1 \\ 2 & 5 \\ 1 & 9 \end{bmatrix}$$

then the transpose of A is

$$A^{T} = \begin{bmatrix} 3 & 2 & 1 \\ -1 & 5 & 9 \end{bmatrix}$$

Transpose of a sum or difference of matrices

The transpose of a sum or difference of matrices is equal to the sum or difference of the transposes of the respective matrices. For example, if

$$A = \begin{bmatrix} 5 & -2 \\ 0 & -1 \end{bmatrix} \qquad B = \begin{bmatrix} 4 & 3 \\ 5 & -6 \end{bmatrix}$$

then

$$(A + B)^{T} = \begin{bmatrix} 9 & 1 \\ 5 & -7 \end{bmatrix}^{T} = \begin{bmatrix} 9 & 5 \\ 1 & -7 \end{bmatrix}.$$

Alternatively,

$$A^T + B^T = \begin{bmatrix} 5 & 0 \\ -2 & -1 \end{bmatrix}^T + \begin{bmatrix} 4 & 5 \\ 3 & -6 \end{bmatrix}^T = \begin{bmatrix} 9 & 5 \\ 1 & -7 \end{bmatrix}.$$

Transpose of a product of matrices

The transpose of a product of matrices is equal to the product of the transpose of the matrices in reverse sequence. Thus if

$$A = \begin{bmatrix} 3 & 0 \\ -4 & -1 \end{bmatrix} \qquad B = \begin{bmatrix} 3 & 5 & -7 \\ 0 & -1 & 8 \end{bmatrix}$$

then

$$AB = \begin{bmatrix} 9 & 15 & -21 \\ -12 & -19 & 20 \end{bmatrix} \quad [AB]^T = \begin{bmatrix} 9 & -12 \\ 15 & -19 \\ -21 & 20 \end{bmatrix}.$$

Alternatively,

$$[AB]^T = B^T A^T = \begin{bmatrix} 3 & 0 \\ 5 & -1 \\ -7 & 8 \end{bmatrix} \begin{bmatrix} 3 & -4 \\ 0 & -1 \end{bmatrix} = \begin{bmatrix} 9 & -12 \\ 15 & -19 \\ -21 & 20 \end{bmatrix}.$$

Special types of matrices

Diagonal matrix

A square matrix (having the same number of rows and columns) that has zeros everywhere except on the main diagonal (the diagonal running from upper left to lower right) is called a diagonal matrix. For example,

$$A = \begin{bmatrix} 2 & 0 \\ 0 & -1 \end{bmatrix} \qquad \text{and} \qquad B = \begin{bmatrix} -3 & 0 & 0 \\ 0 & 0 & 0 \\ 0 & 0 & 1 \end{bmatrix}$$

are diagonal matrices.

Identity matrix

If all the diagonal elements of a diagonal matrix are positive one (+1), then it is called an identity matrix. For example,

$$I = \begin{bmatrix} 1 & 0 & 0 \\ 0 & 1 & 0 \\ 0 & 0 & 1 \end{bmatrix}$$

is the identity matrix of dimension 3, denoted by I_3.

Premultiplying or postmultiplying a matrix by an appropriately sized identity matrix leaves it unchanged. That is,

$$I_m A_{m \times n} = A_{m \times n} I_n = A_{m \times n}.$$

Null matrix

A null matrix, denoted by 0, is a matrix whose elements are all zeros. When an appropriately sized null matrix is added to or subtracted from another matrix that matrix remains unchanged. That is,

$$A_{m \times n} + 0_{m \times n} = A_{m \times n}.$$

Premultipyling or postmultipying a matrix by an appropriately sized null matrix results in another null matrix. That is,

$$A_{m \times n} 0_{n \times p} = 0_{m \times p}.$$

Symmetric matrix

If a square matrix and its transpose are equal, i.e. if $a_{ij} = a_{ji}$ for all i and j, then the matrix is said to be symmetric. For example,

$$A = \begin{bmatrix} 1 & 3 & 5 \\ 3 & 0 & 2 \\ 5 & 2 & 6 \end{bmatrix}$$

is a symmetric matrix.

Idempotent matrix

An idempotent matrix is a symmetric matrix which reproduces itself when it is multiplied by itself. That is, if $A = A^T$ and $AA = A$, then A is an idempotent matrix.

Partitioned matrix

It is sometimes convenient to partition a matrix into submatrices which can be treated as scalars in performing operations, in particular multiplications, on the original matrix. Partitioning of a matrix is indicated by horizontal and vertical broken lines between rows and columns. For example, an $m \times n$ matrix A may be partitioned as

$$A = (A1 \mid A2)$$

where $A1$ is $m \times n_1$, $A2$ is $m \times n_2$ and $n_1 + n_2 = n$.

The transpose of a partitioned matrix can be written in terms of the transposes of its submatrices. Thus,

$$A^{\mathrm{T}} = \begin{bmatrix} A1^{\mathrm{T}} \\ -- \\ A2^{\mathrm{T}} \end{bmatrix}.$$

If two matrices are partitioned comformably, they can be added, subtracted or multiplied. If an $m \times n$ matrix A is partitioned $A = [A1 \mid A2]$ and an $m \times n$ matrix B is partitioned $B = [B1 \mid B2]$, where $A1$ is $m \times n_1$, $A2$ is $m \times n_2$, B_1 is $m \times n_1$ and $B2$ is $m \times n_2$ such that $n_1 + n_2 = n$, then $A + B = [A1 + B1 \mid A2 + B2]$.

The computational advantages of matrix partitioning are associated primarily with multiplication and more complex operations. Note that matrices must be partitioned comformably for multiplication. If $A_{m \times n} = [A1_{m \times n_1} \mid A2_{m \times n_2}]$ where

$$n_1 + n_2 = n, \text{ and } B_{n \times p} = \begin{bmatrix} B1_{n_1 \times p} \\ \overline{B2}_{n_2 \times p} \end{bmatrix}$$

such that $n_1 + n_2 = n$, then

$$AB = [A1 \mid A2] \begin{bmatrix} B1 \\ -- \\ B2 \end{bmatrix} = A1B1 + A2B2.$$

For example, if

$$A = \begin{bmatrix} 1 & 0 & \vdots & -1 \\ 0 & 4 & \vdots & 0 \\ 1 & -1 & \vdots & 2 \end{bmatrix} \quad \text{and } B = \begin{bmatrix} 2 & 1 \\ 0 & 3 \\ -1 & 2 \end{bmatrix}$$

then

$$AB = \underbrace{\begin{bmatrix} 2 & 1 \\ 0 & 12 \\ 2 & -2 \end{bmatrix}}_{A1B1} + \underbrace{\begin{bmatrix} 1 & -2 \\ 0 & 0 \\ -2 & 4 \end{bmatrix}}_{A2B2} = \begin{bmatrix} 3 & -1 \\ 0 & 12 \\ 0 & 2 \end{bmatrix}.$$

The determinant of a matrix

Determinants are defined only for square matrices. The determinant of a matrix is a scalar obtained from the elements of the matrix by specified operations. For example, the determinant of

$$A = \begin{bmatrix} a_{11} & a_{12} \\ a_{21} & a_{22} \end{bmatrix}$$

is $|A| = a_{11}a_{22} - a_{12}a_{21}$.

Determinants of matrices equal to or larger than 3×3 are normally obtained by a procedure known as expansion by cofactors. For example, the determinant of a 3×3 matrix can be obtained as

$$A = a_{11} \begin{vmatrix} a_{22} & a_{23} \\ a_{32} & a_{33} \end{vmatrix} - a_{12} \begin{vmatrix} a_{21} & a_{22} \\ a_{31} & a_{33} \end{vmatrix} + a_{13} \begin{vmatrix} a_{21} & a_{22} \\ a_{31} & a_{32} \end{vmatrix}.$$

Note that each determinant in the sum is the determinant of a sub-matrix of A obtained by deleting a row and a column of A. These determinants are called minors. Thus, if an $(n-1) \times (n-1)$ matrix M_{ij} is obtained by deleting the ith row and jth column of A, then the determinant $|M_{ij}|$ is a minor of the matrix A. The scalar $C_{ij} = (-1)^{i+j}|M_{ij}|$ is called the cofactor.

Thus, the determinant $|A|$ obtained by expansion by cofactors in terms of the ith row can be expressed as

$$|A| = \sum_{j=1}^{m} a_{ij}C_{ij}$$

for any row $i = 1, 2, \ldots n$. If it is expanded in terms of cofactors of the jth column, it is

$$|A| = \sum_{i=1}^{n} a_{ij}C_{ij}$$

for any column $j = 1, 2, \ldots, m$. For computational ease, a determinant should be expanded in terms of the row or column having the largest number of zero elements. For example, the determinant of the matrix

$$A = \begin{bmatrix} 0 & 0 & -2 \\ 6 & -8 & 1 \\ 0 & 3 & 4 \end{bmatrix}$$

should be obtained by expanding in terms of the cofactors of the first row which has the maximum number of zeros. Thus,

$$|A| = 0\begin{vmatrix} -8 & 1 \\ 3 & 4 \end{vmatrix} - 0\begin{vmatrix} 6 & 1 \\ 0 & 4 \end{vmatrix} + (-2)\begin{vmatrix} 6 & -8 \\ 0 & 3 \end{vmatrix}$$

$$= -2[6 \times 3 - (-8)\,0] = -36.$$

The transpose of a matrix of cofactors is called the adjoint of the original matrix from which cofactors are obtained.

The inverse of a matrix

Inverses can be obtained only for square matrices. It is an operation analogous to division in ordinary algebra, i.e. $x/y = (x)(1/y) = xy^{-1}$. However, while every non-zero number has an inverse or reciprocal, there are square matrices, in addition to the null matrix, which do not have inverses and they are known as singular matrices. The determinant of a singular matrix is zero. The matrix which has an inverse is said to be non-singular.

Only one method of matrix inversion is discussed here. It uses adjoints and determinants of the matrix to be inverted. If A is non-singular, i.e. $|A| \neq 0$, then

$$A^{-1} = \frac{1}{|A|} \text{ adjoint } A$$

For example, let

$$A = \begin{bmatrix} 1 & 2 & 3 \\ -1 & 0 & 4 \\ 0 & 2 & 2 \end{bmatrix}$$

$$|A| = 0 + 0 - 6 - (0 + 8 - 4) = -10.$$

We obtain the cofactors as follows:

$$C_{11} = (-1)^{1+1}\begin{vmatrix} 0 & 4 \\ 2 & 2 \end{vmatrix} = -8$$

$$C_{12} = (-1)^{1+2}\begin{vmatrix} -1 & 4 \\ 0 & 2 \end{vmatrix} = 2$$

$$C_{13} = (-1)^{1+3}\begin{vmatrix} -1 & 0 \\ 0 & 2 \end{vmatrix} = -2$$

$$C_{21} = (-1)^{2+1} \begin{vmatrix} 2 & 3 \\ 2 & 2 \end{vmatrix} = 2$$

$$C_{22} = (-1)^{2+2} \begin{vmatrix} 1 & 3 \\ 0 & 2 \end{vmatrix} = 2$$

$$C_{23} = (-1)^{2+3} \begin{vmatrix} 1 & 2 \\ 0 & 2 \end{vmatrix} = -2$$

$$C_{31} = (-1)^{3+1} \begin{vmatrix} 2 & 3 \\ 0 & 4 \end{vmatrix} = 8$$

$$C_{32} = (-1)^{3+2} \begin{vmatrix} 1 & 3 \\ -1 & 4 \end{vmatrix} = -7$$

$$C_{33} = (-1)^{3+3} \begin{vmatrix} 1 & 2 \\ -1 & 0 \end{vmatrix} = 2.$$

Therefore, the cofactor matrix is

$$C = \begin{bmatrix} -8 & 2 & -2 \\ 2 & 2 & -2 \\ 8 & -7 & 2 \end{bmatrix}$$

$$\text{adjoint } A = C^{T} = \begin{bmatrix} -8 & 2 & 8 \\ 2 & 2 & 7 \\ -2 & -2 & 2 \end{bmatrix}.$$

Thus

$$A^{-1} = \frac{1}{|A|} \text{ adjoint } A = \begin{bmatrix} \frac{4}{5} & -\frac{1}{5} & -\frac{4}{5} \\ -\frac{1}{5} & -\frac{1}{5} & \frac{7}{10} \\ \frac{1}{5} & \frac{1}{5} & -\frac{1}{5} \end{bmatrix}.$$

Unless there is a good reason to believe that the matrix is non-singular, it is advisable to find its determinant first. If the determinant is found to be zero then a lot of useless computation can be avoided. One easy way to discover whether a determinant is zero is to check whether two rows or columns are identical or a multiple (linear transformation) of each other. If this is the case then the determinant is zero.

Inversion of a partitioned matrix

It is sometimes convenient to obtain the inverse of a matrix in partitioned form. If an $n \times n$ matrix A is partitioned as

$$A = \begin{bmatrix} A11 & A12 \\ \hline A21 & A22 \end{bmatrix}$$

where $A11$ is $n_1 \times n_1$, $A12$ is $n_1 \times n_2$, $A21$ is $n_2 \times n_1$, A_{22} is $n_2 \times n_2$ and $n_1 + n_2 = n$, then

$$A^{-1} = \begin{bmatrix} A11^{-1}[I + A12B^{-1}A21A11^{-1}] & -A11^{-1}A12B^{-1} \\ -B^{-1}A21A11^{-1} & B^{-1} \end{bmatrix}$$

where $B = [A22 - A21 \; A11^{-1} \; A12]$ and $A11$ and B are non-singular.

For example, let A be partitioned as

$$A = \begin{bmatrix} -1 & 2 & 3 \\ \hline -1 & 0 & 4 \\ 0 & 2 & 2 \end{bmatrix} = \begin{bmatrix} A11 & A22 \\ \hline A21 & A22 \end{bmatrix}.$$

Then

$$B = [A22 - A21 \; A11^{-1} \; A12]$$

$$= \begin{bmatrix} 0 & 4 \\ 2 & 2 \end{bmatrix} - \begin{bmatrix} -1 \\ 0 \end{bmatrix} [1][2 \quad 3]$$

$$= \begin{bmatrix} 2 & 7 \\ 2 & 2 \end{bmatrix}$$

and

$$B^{-1} = \begin{bmatrix} -\frac{1}{5} & \frac{7}{5} \\ \frac{1}{5} & -\frac{1}{5} \end{bmatrix}$$

$$-A11^{-1}A12B^{-1} = [-1] \; [2 \quad 3] \begin{bmatrix} -\frac{1}{5} & \frac{7}{10} \\ \frac{1}{5} & -\frac{1}{5} \end{bmatrix}$$

$$= [-\frac{1}{5} \quad -\frac{4}{5}]$$

$$-B^{-1}A21A11^{-1} = \begin{bmatrix} \frac{1}{5} & -\frac{7}{5} \\ -\frac{1}{5} & \frac{1}{5} \end{bmatrix} \begin{bmatrix} -1 \\ 0 \end{bmatrix} [1]$$

$$= \begin{bmatrix} -\frac{1}{5} \\ \frac{1}{5} \end{bmatrix}$$

and so,

$A11^{-1}[I + A12B^{-1}A_{21}A11^{-1}]$

$$= [1]\left([1] + [2 \quad 3]\begin{bmatrix} -\frac{1}{5} & \frac{7}{10} \\ \frac{1}{5} & -\frac{1}{5} \end{bmatrix}\begin{bmatrix} -1 \\ 0 \end{bmatrix} [1]\right)$$

$$= [1]\left([1] + [\frac{1}{5} \quad \frac{4}{5}]\begin{bmatrix} -1 \\ 0 \end{bmatrix}\right)$$

$$= 1(1 - \frac{1}{5}) = \frac{4}{5}.$$

Therefore,

$$A^{-1}\begin{bmatrix} \frac{4}{5} & -\frac{1}{5} & -\frac{4}{5} \\ -\frac{1}{5} & -\frac{1}{5} & \frac{7}{10} \\ \frac{1}{5} & \frac{1}{5} & -\frac{1}{5} \end{bmatrix}.$$

Properties of inverses

1 The inverse of the inverse of a matrix is the original matrix, i.e. $[A^{-1}]^{-1} = A$.
2 The determinant of the inverse of a matrix is equal to the reciprocal of the determinant of the matrix, i.e. $|A^{-1}| = 1/|A|$
3 The inverse of the transpose of a matrix is equal to the transpose of the inverse of the matrix, i.e. $[A^T]^{-1} = [A^{-1}]^T$.
4 The inverse of the product of two matrices is equal to the product of their inverses in reverse order, i.e. $[AB]^{-1} = B^{-1}A^{-1}$.

4

MULTI-SECTORAL MODELS AND THE SOCIAL ACCOUNTING MATRIX

INTRODUCTION

The inability of the input–output model to deal with issues of income distribution and growth and the incomplete nature of multiplier analysis can be overcome by extending the input–output table to include institutional accounts – income and expenditure of households and other socio-economic groups. In terms of the accounting framework, it involves the mapping of factoral incomes into incomes of various households and institutions. This extended data system with institutional accounts in it is known as the social accounting matrix (SAM). The term 'social' as opposed to 'national' has special significance which arises from the attempt to classify various institutions according to their socio-economic background (e.g. rural–urban, poor–rich, etc.) instead of their economic or functional activities (e.g. labour–capitalist) alone, as in national and input–output accounts. Since a SAM contains both social and economic data, it provides a conceptual basis to examine growth and distributional issues within a single analytical framework. This is a distinct advantage over the traditional national and input–output accounts which record 'economic' transactions alone irrespective of the social background of the transactors.

THE SOCIAL ACCOUNTING MATRIX

To explain how a SAM can be used to analyse income distribution as an integral part of the growth process we have presented a SAM for a country in a schematic form in Table 4.1.

The whole arrangement can be thought of as a partitioned matrix. That is, each element T_{ij} in Table 4.1 represents a submatrix. Thus,

Table 4.1 A Schematic SAM

	Expenditure				
Receipts	*Factors of production*	*Institutions*	*Production activities*	*Other accounts*	*Total*
Factors of production	0	0	T_{13}	T_{14}	t_1
Institutions	T_{21}	T_{22}	0	T_{24}	t_2
Production activities	0	T_{32}	T_{33}	T_{34}	t_3
Other accounts	0	T_{42}	T_{43}	T_{44}	t_4
Total	$\hat{t}_1$	$\hat{t}_2$	$\hat{t}_3$	$\hat{t}_4$	

T_{33} is the input–output flow matrix which records transactions between producing sectors or industries. When written explicitly,

$$T_{33} = \begin{bmatrix} X_{11} & X_{12} & \dots & X_{1n} \\ X_{21} & X_{22} & \dots & X_{2n} \\ \vdots & & \ddots & \vdots \\ X_{n1} & X_{n2} & \dots & X_{nn} \end{bmatrix}$$

where the X_{ij} represent sales from the ith sector to the jth sector. Similarly, the submatrix T_{13} represents payments by producing sectors to factors of production (link 2 in terms of the flow diagram in Figure 3.2), and T_{21} maps household and other institutional incomes from factor incomes (link 3 of the flow diagram). The transformation of institutional incomes into demand (expenditure) for production activities (link 1) is recorded in the submatrix T_{32}. The row sum t_i is the total receipts of the ith account and the column sum $\hat{t}_i$ represents the total expenditure. By the accounting rule, $t_i = \hat{t}_i$. On the assumption that each account distributes its income in fixed proportions we can normalize each transaction submatrix T_{ij} by the vector $\hat{t}_j$ to give

$$A_{ij} = T_{ij}\hat{t}_j^{-1}. \tag{4.1}$$

That is, A_{ij} is obtained by dividing each element of T_{ij} by the sum of the column in which it appears. Assuming that there are only two types of household (namely, rural and urban) and two production activities (namely, agricultural and manufacturing), the normalized coefficient submatrix for the institution accounts, for example, will be as follows:

$$A_{32} = T_{32}\hat{t}^{-1} = \begin{vmatrix} C_{ar}/C_r & C_{au}/C_u \\ C_{mr}/C_r & C_{mu}/C_u \end{vmatrix}$$

where C_i is the total consumption expenditure of the ith type of household (r = rural, u = urban) and C_{ki} is the consumption of the kth sector (a = agriculture, m = manufacturing) by the ith household type. Since total outlay is equal to total income, C_{ki}/C_i represents the ith type household's average propensity to consume (spend) the kth sector's product. A similar interpretation can be given to all other A_{ij}s.

In terms of partitioned matrices of average expenditure propensities, we can, therefore, write the balance equation as

$$\begin{bmatrix} t_1 \\ t_2 \\ t_3 \\ t_4 \end{bmatrix} = \begin{bmatrix} 0 & 0 & A_{13} \\ A_{21} & A_{22} & 0 \\ 0 & A_{32} & A_{33} \\ 0 & A_{42} & A_{43} \end{bmatrix} \begin{bmatrix} \hat{t}_1 \\ \hat{t}_2 \\ \hat{t}_3 \end{bmatrix} + \begin{bmatrix} Q_1 \\ Q_2 \\ Q_3 \\ Q_4 \end{bmatrix}$$

where Q_i is the vector of row sums of the submatrix T_{i4} for i = 1, 2, 3, 4. Assuming that each of the Q_is is an exogenous set of numbers we can decompose the above expressions as

$$\begin{bmatrix} t_1 \\ t_2 \\ t_3 \end{bmatrix} = \begin{bmatrix} 0 & 0 & A_{13} \\ A_{21} & A_{22} & 0 \\ 0 & A_{32} & A_{33} \end{bmatrix} \begin{bmatrix} \hat{t}_1 \\ \hat{t}_2 \\ \hat{t}_3 \end{bmatrix} + \begin{bmatrix} Q_1 \\ Q_2 \\ Q_3 \end{bmatrix} \qquad (4.2)$$

and

$$t_4 = A_{42}\hat{t}_2 + A_{43}\hat{t}_3 + Q_4. \qquad (4.3)$$

The above decomposition implies that we can derive the t_4 set of accounts once we have obtained the other three sets of accounts. Therefore, we shall concentrate only on (4.2) which gives us balance equations for factors, institutions and production accounts. Since $t = \hat{t}$, we can express (4.2) in matrix form as

$$t = At + Q. \qquad (4.2a)$$

After necessary manipulations, we can obtain

$$t = [I - A]^{-1}Q. \qquad (4.4)$$

This is an analogous procedure to that followed for solving the conventional input–output model. But in this case the inverse $[I - A]^{-1}$

is more generalized as it includes the feedback from income generation to demand and as such provides 'Keynesian' multipliers for various producing sectors. This becomes clear if we write the production account (which is the same as the input–output account) explicitly. In terms of equations system (4.2), this implies solving for sectoral output submatrix t_3 as

$$t_3 = [I - A_{33}]^{-1}(A_{32}t_2 + Q_3). \tag{4.5}$$

As explained before, A_{32} represents the marginal propensities to spend by the households on different sectoral products. That is, the coefficient submatrix A_{32} relates expenditure to institutional incomes and determines induced demand. t_2 is the level and distribution of income across institutions. Therefore, $[I - A_{33}]^{-1}A_{32}t_2$ gives the sectoral distribution of induced demand. In the conventional input–output analysis, final demands are treated as exogenous and hence A_{32} is zero. Therefore, in terms of equations system (4.5), the conventional input–output solution for sectoral output level will be

$$t_3 = [I - A_{33}]^{-1}Q_3. \tag{4.6}$$

Owing to the absence of induced expenditure coefficients A_{32} in (4.6), the simple input–output model misses the link between production activities and induced demand and thereby short-circuits the 'Keynesian' multiplier process. As a result, it fails to address the question of absorptive capacity. Hence, conventional planning based on simple input–output forecasting relies heavily on public spending to take up any slack in demand. In addition, the conventional input–output model cannot explain the production structure as determined by the initial income distribution.

On the other hand, a SAM-based model as presented in the equations system (4.2) describes a system in which the production structure and income distribution are determined simultaneously. Therefore, it can be used to formulate a strategy of growth with distribution. The solution of (4.2) for sectoral output levels like equations (4.5) can be used to measure the impact of exogenously introduced income distribution measures (i.e. changes in t_2) on production structure and growth. For example, in 1973 the Indian Planning Commission estimated that an income distribution which would bring the poorest third of the population up to their level of 'basic needs' might change output growth rates in twenty-two of the sixty-six sectors by more than 1 per cent (Government of India (1973); see also Cline (1972) for similar exercises on Latin America).

Income redistribution changes the sectoral output rates because different income and social groups have different marginal propensities to spend on the product of each sector. Thus, a change in the income distribution generates a new demand pattern which, in turn, gives rise to a new structure of production. A reduction in inequality, in general, is likely to raise demand for necessities and reduce consumption demand for luxury goods. To the extent that the wage goods and necessities are produced in the informal sector and use more labour intensive methods, the new demand pattern and production structure will increase employment and further improve income distribution. It also follows that a SAM-based model like (4.2) can be used to examine probable impacts of exogenously introduced sectoral investment programmes on income across institutions. This is because different producing sectors use labour in different proportions. Since employment is the single most important source of income, different growth rates of employment (due to different sectoral growth) will generate different income distributions (see Maton and Joos (1984) for such exercises).

However, there is another and more controversial issue in the growth with redistribution strategy. While few economists would deny the link between income distribution and production structure via the demand pattern, there are disagreements regarding the supply side of the problem. The view of most economists is that the rich are likely to save more, relative to their income, than the poor. Therefore, it is believed that the shift of relative incomes and wealth in favour of the poor will reduce the total amount of domestic savings and, other things being equal, this will reduce the amount of investment and the economy's expansion rate. There are some economists who dispute this view, however, and hold that the rich in less developing countries (LDCs) are not high savers. Rather they indulge in 'conspicuous consumption'. They spend on imported luxury products which strains the country's precarious foreign exchange constraint. Even if the rich in LDCs save more, their savings are mostly locked up in unproductive forms such as gold, jewellery or landed property.

This is one of those disputes in economics which cannot be settled without reference to evidence. The SAM can go a long way in shedding light on this empirical issue. The accounts for institutions in a SAM are subdivided into (i) current account and (ii) capital account. The current account records consumption by various socio-economic groups and it allows us to check the validity of the contention that the rich in LDCs indulge in conspicuous consumption or spend disproportionately more on imported products. The current and capital accounts are interlinked through savings. The difference between income and consumption (i.e. savings) enters into the capital account as a receipt. On the expenditure side the capital account shows who invests on what. Thus the capital account in a SAM framework allows us to examine both savings and investment behaviours of different socio-economic groups.

Assuming that the rich in a particular country save more and invest productively, it is still possible to design an appropriate strategy which can harmonize the concern for growth and poverty eradication. First, it is necessary to distinguish between poverty and inequality. Experience shows that a reduction of relative inequality does not necessarily eradicate pockets of poverty among certain socio-economic groups. Poverty among target groups can perhaps be best tackled by a 'basic needs' approach. Even though there are disagreements about the appropriate definition of basic needs, the commodity composition of consumption of the target socio-economic groups can give an indication of an absolute level of poverty (Rao 1988). Policies can then be designed to enlarge the supply of 'basic needs' goods and to ensure that the target groups have enough income to obtain such goods while at the same time maintaining some income differentials for those above the 'poverty line' (Pyatt and Thorbecke 1976).

LIMITATIONS OF MULTI-SECTORAL PLANNING MODELS

Despite their numerous uses, the input–output and SAM-based multi-sectoral planning models have certain limitations. These limitations stem from their assumptions regarding technology. Two of the fundamental input–output assumptions are that

1 the technology and capital coefficient matrices are constant and
2 the technology and capital coefficient matrices are unique, i.e. there is only one production technique available to each sector or industry.

There are two problems associated with the first assumption. The assumption of constant technical and capital coefficients implies constant returns to scale and hence appears to be in contradiction with economic development characterized by the phenomenon of increasing returns to scale. Since investment goods are indivisible, a 'big push' in investment is likely to generate increasing returns to scale. In fact, market failure arising from increasing returns to scale is one of the arguments used in favour of planning. Besides capacity expansion, investment planning aims at introducing new methods of production and improved technologies. Therefore, it is extremely hazardous to project output and capital requirements based on existing technology and capital coefficients which are likely to change during the process of economic development owing either to deliberate attempts or to indivisibilities of investment.

Second, one of the main objectives of development planning is to bring about a structural change of the economy. This involves initiating new activities and phasing out some existing ones. The new configuration of activities will therefore generate a different input–output relationship. The existing input–output relationships appear to offer little or no help in considering the establishment of new industries or activities.

The problem of variable coefficients is generally tackled by forecasting changes in the input structure and product mix on the basis of past experience. The most widely used forecasting method is known as RAS and was developed by Professor Richard Stone. The RAS is a mechanical method of updating the intermediate uses to force equality between total uses and industry gross output, information on which is readily available. It bypasses thereby the long and time consuming process of collecting detailed information on intermediate purchases. The problem of introducing new activities is handled by adding new rows and columns with technical coefficients obtained either from engineering data or on the basis of statistical experience of similar activities already established in countries of comparable economic development.

The second assumption of a single process production function poses a more serious problem. It precludes the possibility of substitution and alternative ways of doing things. In reality, though, inputs can be combined in various proportions to obtain the same level of output. That is, there exists more than one set of technical coefficients. The job of the economist is to choose the most economical one from the set of technically efficient production processes. In technical terms, the problem is to choose the optimal production process, given the scarcity price of resources under their existing configuration.

The standard input–output technique which is the core of multi-sectoral planning models cannot handle the problem of choosing the optimal production process which would yield the same total output while making the least use of whatever resources are most limited. This involves supplementing the standard input–output methodology with the linear programming method. This is the topic which we will take up in the next chapter.

FURTHER READING

The construction and uses of SAMs in developing countries is examined in Bulmer-Thomas (1982). Taylor (1979) provides examples of the application of the SAM to a number of developing economies. Pyatt *et al.* (1977) give a detailed description of the application of the social accounting approach to Sri Lanka; the book also contains a very useful introductory chapter by Stone summarizing the main characteristics of the SAM technique.

5

PROGRAMMING APPROACH TO PLANNING

INTRODUCTION

The models that we have discussed so far have been concerned with the achievement of certain targets. These targets are set arbitrarily in the sense that they are not derived from any optimization criteria. As such these consistency models do not incorporate any explicit national objective function to be maximized (or minimized) subject to constraints. But as we know, economics is a science of choice within constraints. No economy can produce as much as it wants. At any point in time it is bound by resource availability (or the production possibility frontier). Over the course of time, the production possibility frontier is likely to expand; but that does not eliminate the problem of choice. The planners' job is to find an 'optimum' allocation of resources in the short to medium term, keeping in view the long-run objective of expanding the economy's production possibility frontier.

The absence of any optimization criteria in the consistency models has two consequences. First, it prevents the planners from examining an economy's production possibilities, given the existing and the likely future configuration of resources. That is, it is not possible to know whether a target level and combination of outputs is really on the production possibility frontier and whether it is the 'best' combination, given the society's value judgement. In short, the consistency models do not address the question of 'optimum' allocation of resources. The second consequence, which follows from the first, is that the consistency models fail to consider the efficiency in the use of resources by minimizing the cost.

The duality between 'optimum' allocation and 'efficient' use of resources can be explained by using a simple analogy from the theory of production in micro-economics. As we know, the maximization of output subject to a given cost is obtained at a point where the iso-cost line is tangent to an isoquant. The same optimum point also represents the minimum cost of producing that particular level of output. The situation is depicted in Figure 5.1. Point E represents the maximum level of output obtainable at a cost given by AB. It can also be interpreted as a point on the minimum cost AB line to produce a fixed level of output given by the isoquant to which AB is tangent. Thus, the mirror image of the 'optimum' allocation of resources is the 'efficient' use of resources. At the optimum point, the marginal rate of technical substitution between resources (or marginal rate of product transformation) is equal to the relative prices of resources. These relative prices are efficiency prices in the sense that they satisfy the 'Pareto optimality' criterion of price equals marginal cost (this point is taken up in a later section).

Since one of the main tasks of development strategy is to ensure the optimum allocation and efficient use of resources, it is necessary to extend the consistency models into programming models. The programming models provide a simultaneous solution to the optimum allocation of resources in terms of objective functions and efficiency in the use of resources through the proper valuation of scarce resources and the avoidance of social waste. To quote Dorfman (1953: 810), one of the early contributors to the development of the programming method, 'The most useful applications of mathematical programming are probably to problems concerned with finding optimal production plans using specified quantities of some or all of the resources involved.' The programming approach to resource allocation begins

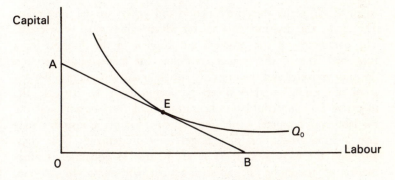

Figure 5.1 Maximization of output/minimization of cost

with the problem of balancing supply and demand for different commodities and factors of production. According to Chenery (1961: 32),

> Historically speaking, the programming approach is thus the operational counterpart of the theory of balanced growth. It is only with the development of linear programming that it is possible to reconcile the consistency criteria and the productivity criteria in a systematic way. A link between the test of consistency (feasibility) in resource allocation and the test of productivity (efficiency) is provided by a consideration of the price implications of a given allocation.

Concepts of mathematical programming

Programming addresses optimization problems with a specific structure: maximize (or minimize) an objective function subject to a set of constraints which define feasibility. The objective function and constraints include decision variables and parameters. Decision variables are controllable while parameters are given. The general form of a mathematical programming problem can be written as

$$
\begin{aligned}
\text{maximize (or minimize)} \quad & Z(X_1, X_2, \ldots, X_n) = 0 \\
\text{subject to} \quad & g_1(X_1, X_2, \ldots, X_n) = 0 \\
& g_2(X_1, X_2, \ldots, X_n) = 0 \\
& \qquad \vdots \qquad \vdots \\
& g_n(X_1, X_2, \ldots, X_n) = 0
\end{aligned}
\tag{5.1}
$$

where $Z(X_i)$ is the objective function and $g_i(X_i)$ are the constraints. The X_i are the decision variables for which values must be selected. The parameters of the problem are implicit in the symbols for a function. That is, the parameters define the relationships between decision variables and the objective function and constraints. We can think of decision variables as sectoral output levels. In that case, the objective function could be maximization of total output and the availability of resources would define the constraints. Constraints may also include satisfaction of a target level of demand. Note that, although the constraints in (5.1) are written as equalities, mathematical programming does admit inequality constraints. In this case, the constraints would mean that the total use of a particular resource must not exceed its availability. If constraints with a target demand are included in the problem, an inequality relation would mean that the supply of output must at least satisfy a minimum (or target) level of demand. Thus, the recognition of inequality relations allows the possibility of slacks in the use of (relatively abundant) resources and production beyond the basic minimum requirement. The possibility of slacks in the resource use has an important implication in terms of opportunity cost and efficiency (or shadow) pricing which will be discussed in the later part of this chapter.

In structuring and solving a mathematical programme, the analyst attempts to discover decisions about the system under study that are in some sense 'best'. Thus, a solution to the programming problem entails the collection of values for each of the decision variables. A solution which satisfies all the constraints is known as a 'feasible' solution. In practice there will be an infinite number of feasible solutions which together map an economy's feasibility frontier (or production possibility frontier). The role of an objective function is to provide a basis for the evaluation of the feasible solutions. That is, with the value judgement implicit in the objective function the analyst can compare various combinations of outcome on the feasibility frontier and choose the 'best' one.

LINEAR PROGRAMMING

The most widely used programming method is linear programming (LP). LP is a special case of mathematical programming in which all the functions in (5.1) are assumed to be linear. Linearity implies that the objective function and all the constraints are summations of decision variables, each of which is multiplied by a coefficient or parameter. That is, variables are in additive relation in each equation, and no variable is raised to a power higher than unity. One further assumption is that the decision variables must be non-negative. For the purpose of illustration, we shall formulate a simple LP problem in this section.

Let us assume that a country produces two products: agricultural and manufactured. Each unit of agricultural product sells at $1 and the manufactured product at $4. It takes two workers and four units of capital to produce one unit of agricultural product and ten workers and five units of capital to produce one unit of manufactured product. The total number of workers and amount of capital available are 1,000 and 600 respectively. We shall structure a linear programme to allocate the country's resources (labour and capital) so as to maximize the total value of output or GDP.

The first step is to define the decision variables and parameters. The decisions are the amounts of products (levels of agricultural and manufactured output) that the country should produce. Let us define the following: X_1 is the level of agricultural product and X_2 is the level of manufactured product. The parameters of the problem are as follows:

1 number of workers required per unit of agricultural and manufactured products;
2 units of capital required per unit of agricultural and manufactured products;
3 total available amount of resources (labour and capital);
4 price per unit of agricultural and manufactured products.

The next step is to state the objective function and constraints. The objective is to maximize the value of total output (or GDP). Therefore the objective function is

$$\text{maximize} \quad Z \quad = 1X_1 \quad + 4X_2.$$
$$\text{(total value} \quad \text{(value of} \quad \text{(value of}$$
$$\text{of output)} \quad \text{agriculture)} \quad \text{manufacture)}$$

Note that the units of Z and the two terms on the right-hand side of the objective function are the same. That is, they are expressed in value or dollar terms.

The constraints are on the availability of resources. That is, in producing the two products we cannot use more of any resource than is available. Thus, the constraints take the form

$$\text{resources used} \leqslant \text{resources available.}$$

Therefore, in terms of the parameters and decision variables, the constraints are

$$2X_1 + 10X_2 \leqslant 1,000 \quad \text{(labour constraint)}$$
$$4X_1 + 5X_2 \leqslant 600 \quad \text{(capital investment).}$$

The condition that negative amounts of neither product can be produced implies a further constraint as

$$X_1 \geqslant 0 \quad X_2 \geqslant 0.$$

Therefore, we can summarize the country's resource allocation problem as

$$\text{maximize} \quad Z = 1X_1 + 4X_2$$
$$\text{subject to} \quad 2X_1 + 10X_2 \leqslant 1{,}000$$
$$4X_1 + 5X_2 \leqslant 600$$

with non-negativity conditions

$$X_1 \geqslant 0 \qquad X_2 \geqslant 0.$$

The solution to the above LP problem entails finding values for X_1 and X_2 which will maximize Z and at the same time are feasible in the sense that the attainment of these values does not violate any of the constraints (including the non-negativity conditions).

Solution of linear programming problems

The set of feasible solutions to a linear programme can be represented graphically when there are not more than three decision variables. If the number of decision variables exceeds three then the solution is obtained by using a method known as 'simplex'. The simplex method is discussed in the appendix to this chapter. However, since our hypothetical LP problem involves only two decision variables, it can be solved by using the graphical method.

The first step in the graphical solution is to convert the inequality constraints into equality constraints so that they can be drawn as straight lines (recall the straight line budget constraint in the consumer's utility maximization problem). Since '$\leqslant$' type inequalities imply the possibility of slacks in the resource use, they can be converted into linear equalities by introducing slacks explicitly in the constraints. Thus, if we denote unused labour by S_1 and unused capital by S_2 then the constraints of our LP problem will be

$$2X_1 + 10X_2 + S_1 = 1{,}000$$
$$4X_1 + 5X_2 + S_2 = 600.$$

But this means we now have four decision variables (two choice variables X_1, X_2; and two slack variables S_1, S_2) and we can no longer use the graphical method of solution. In order to restrict ourselves to two decision variables so that we can use the graphical method, we shall assume that all the resources are fully utilized such that $S_1 = S_2 = 0$. Having done so, we can rewrite our LP problem as

$$\text{maximize} \quad Z \quad = 1X_1 + 4X_2$$
$$\text{subject to} \quad 2X_1 + 10X_2 = 1{,}000$$
$$4X_1 + 5X_2 = 600$$
$$X_1 \geqslant 0 \quad X_2 \geqslant 0.$$

Figure 5.2 provides an illustration of our hypothetical LP problem. The first decision variable, the level of agriculture output X_1, is represented along the horizontal axis and the second decision variable, the level of manufactured output X_2, is represented along the vertical axis. Each constraint is represented by a straight line – AB for labour and CD for capital. The non-negativity conditions are satisfied along the axes. Points to the right of AB violate the labour constraint in the sense that attainment of output levels represented by such points would require more labour than is available. Similarly, output levels to the right of CD would require more capital than is available.

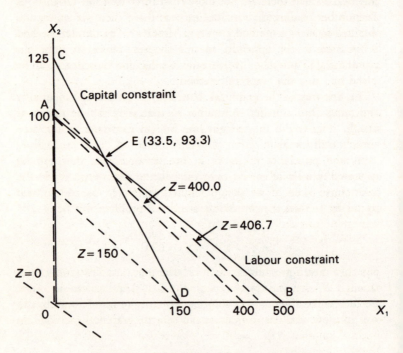

Figure 5.2 A hypothetical linear programme

Therefore, to be feasible, solutions must lie on or on the appropriate side (in this case, to the left) of each of the lines defining resource constraints. OAED, then, represents the feasible region as output levels given by points on or inside the boundary are attainable without violating any of the constraints. On the other hand, CAE and BED are infeasible regions as they violate the labour and capital constraints respectively.

The objective function can be represented in the plot of the feasible region as a series of parallel lines, each of which is a contour that shows combinations of the decision variables that give a given fixed value of the objective function. The value of the function increases as we move from one contour to another in a 'north easterly' direction. Thus, these contours are similar to an isoquant map.

Four objective function contours for $Z = 0$, 150, 400 and 406.7 are drawn in Figure 5.2, although there are an infinite number of them. A combination of values for X_1 and X_2, (i.e. a solution) that lies on a contour would give a GDP equal to the value of the objective function on that contour. Thus, they can be termed 'iso-GDP' lines. The optimal solution to an optimization problem is that feasible solution which yields the highest value of the objective function. That is, we should look for the highest objective function contour that passes through at least one feasible point. In Figure 5.2 such a contour is the line $Z = 406.7$ which passes through the feasible point (33.5, 93.3). Therefore, if the country in the example produces 33.5 units of agricultural product and 93.3 units of manufactured product, it will utilize all its available resources (i.e. $S_1 = 0$, $S_2 = 0$) and maximizes the value of its GDP.

As mentioned earlier, graphical methods of solution are applicable to problems with only two or at most three decision variables. For the solutions to problems with more than three decision variables we need to use the simplex method. However, the simplex method uses some important properties of the graphical solution. The first point to notice is that the optimal solution will always be on the boundary of the feasible region, because an interior point must always lie on an objective function contour that is lower than the contours passing through at least one point on the boundary. That is, one can always find a feasible direction in which to move from an interior point that will improve the objective function. The second point is that the optimal solution is at the 'corner' where two constraints (boundary lines) intersect.

From the above observations of the optimal solution a basic theorem of linear programming is derived. It states that the number of variables (decision and slack) with non-zero-valued solutions will be equal to the number of constraints. In our example, the two intersecting constraints are labour and capital availability, implying the full utilization of both and hence $S_1 = 0$, $S_2 = 0$. Therefore, X_1 and X_2 have non-zero-valued solutions. It is also possible to have a solution at the corner where one of the resource constraints intersects an axis (e.g. points A and D). If, for example, the optimal solutions were at A where the labour constraint and the vertical axis (representing manufactured output) intersect, the two non-zero-valued variables would be X_2 and S_2. $S_2 > 0$ implies that the non-intersecting resources (capital) constraint is not binding and it will not be fully utilized. At the corner point 0, neither of the resource constraints is binding and both the resources will be completely unutilized, implying zero values for the output levels (i.e. $S_1 = 1,000$, $S_2 = 600$, $X_1 = 0$ and $X_2 = 0$).

The basic theorem of LP has an important implication which is that in solving LP problems we can restrict our attention to extreme points of the feasible solution. These extreme points are known as 'basic' solutions. Since the origin is a basic solution, for maximization problems we can start from it and move in the feasible direction of improvement. Algebraically it amounts to substituting each of the basic solutions into the objective function and the one which yields the highest value for the objective function is the optimal solution. The simplex method is based on this basic principle.

Linear programming and the choice of technique

In our previous example, only one production technique or process for each product was considered. The problem was to find the optimal sectoral output levels to be produced by combining inputs (labour and capital) in a given way. It has not sought to examine whether the technique (input ratios for each product) itself is optimal. The assumption of a single process production function appears to be unrealistic. As a matter of fact, inputs can be combined in more than one way to obtain the same level of output. It has been mentioned earlier that one of the tasks of a planner is to choose from available techniques the best one in the sense of minimizing costs and using the least amount of the most scarce resource. Linear programming is a useful tool for the choice of technique.

Let us assume that the country in our previous example can produce the manufactured product in two alternative ways. The processes are as follows:

> process 1: 10 labour and 5 capital
> process 2: 5 labour and 10 capital.

Process 1 is the same as in our previous example. That is, it requires ten units of labour and five units of capital to produce each unit of manufactured product. On the other hand, to produce the same one unit of output process 2 uses five units of labour and ten units of capital. Therefore, process 2 is more capital intensive than process 1.

The fact that each process uses inputs in a fixed proportion implies constant returns to scale, and each process can be represented in the input space as a ray passing through the origin. Process 1 is therefore represented by ray 1 and process 2 by ray 2 in Figure 5.3. Since the same output level can be achieved using either process, let us assume that the output levels given by point A on ray 1 and point B on ray 2 are the same. The line segment containing A and B can be considered an isoquant, for the same output can be produced using some combination of the two processes.

Assuming that the objective is to minimize the cost of producing a fixed level of output given by AB, we have introduced a single linear

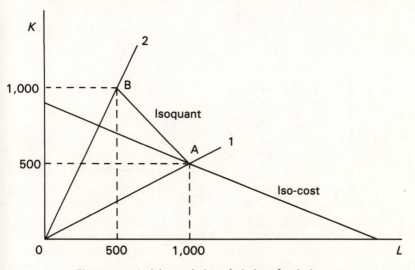

Figure 5.3 Activity analysis and choice of techniques

iso-cost or price line in Figure 5.3. The slope of this iso-cost shows the relative price of the two factor inputs. The flatter slope indicates that labour is less expensive than capital. The feasible direction of movement will be towards the origin in the sense that it will imply lower cost. As can be seen from Figure 5.3, the exclusive use of the labour intensive process 1 – ten labour and five capital – would be the least cost method of producing a given level of output.

The optimal solution will be affected if, with the inclusion of the factor-limitation line as a constraint, the 'solution' now lies in the non-feasible region. This is illustrated in Figure 5.4. The vertical line MN at L = 800 defines the factor-limitation constraint for labour. As can be seen, the previous 'optimal' solution (A) now is unobtainable as it lies to the right of MN. There is an insufficient quantity of labour to rely solely on process 1. The optimal solution will now be at point C which represents a combination of the two processes. Indeed with two constraints – a factor limitation and a fixed output level – the optimal (least cost) feasible solution would require some combination of the two processes. The particular combination of the two processes implied by C may be found by drawing a line from C to D parallel

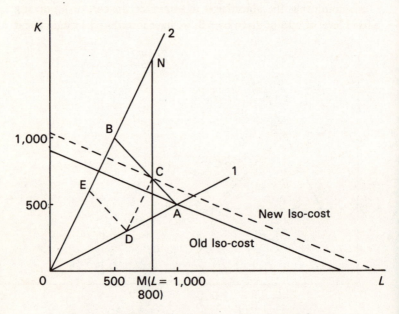

Figure 5.4 Choice of techniques with a limited factor

to process 2. CD (= EB) represents the output produced by process 2 and OD the output produced by process 1. Alternatively, we can draw a line from C to OB and parallel to OA and find the breakdown of output produced by using processes 1 and 2.

SHADOW PRICING OF RESOURCES

While we have introduced relative prices of resources in the problem of choice of techniques, we have not asked whether those prices reflected 'true' opportunity costs. The existence of market distortions in LDCs means that the observed market prices often do not measure 'true' economic or opportunity costs. But in allocating scarce resources to competing uses it is necessary to compare the returns from them with the opportunity costs of resources. As pointed out by Tinbergen (1955), Frisch (1958) and Chenery (1955), if market prices in LDCs cannot be relied upon for evaluating marginal contributions and costs of resources, investment projects should be appraised by using 'shadow' or 'accounting' prices. These are the prices 'that would prevail if (i) the investment pattern under discussion were actually carried out, and (ii) (competitive) equilibrium existed on the markets' (Tinbergen 1958: 39, parentheses added). The use of shadow prices in evaluating investment projects in LDCs is now a well-accepted planning procedure and the details of the methodology of shadow pricing are discussed in Chapter 7.

As mentioned earlier, the programming method provides a simultaneous solution to the optimum allocation and valuation of resources. This follows from the fact that cost minimization is the mirror image of output maximization. Therefore, the LP that maximizes national output also finds prices of resources that minimize the cost of producing a combination of outputs and these prices can be treated as shadow or accounting prices. In this section, we shall illustrate the relationship between optimal allocation and efficiency (shadow) pricing solutions.

The original linear programme is usually referred to as a 'primal' programme and its mirror image as a 'dual' programme. The dual programme bears certain relationships to the primal programme. These relationships are summarized below.

1 The dual of a maximization problem is a minimization problem, and vice versa.

2 The dual has one decision variable for each constraint in the primal problem.
3 The dual has as many constraints as there are decision variables in the primal problem.
4 Coefficients in the primal objective function appear as constants (parameters) in the dual constraints, and vice versa.
5 Each column of coefficients in the primal constraints becomes a row of coefficients in the dual constraints.
6 Inequality signs in the primal constraints are reversed in the dual constraints.
7 The non-negativity conditions ($\geqslant 0$) on decision and slack variables are retained.

Following the above relationships between a primal programme and its dual, we can convert our resource allocation problem to maximize national output into a cost minimization or efficiency pricing problem as follows.

Primal:	maximize	$Z = 1X_1 + 4X_2$
	subject to	$2X_1 + 10X_2 \leqslant 1,000$
		$4X_1 + 5X_2 \leqslant 600$
	non-negativity	$X_1 \geqslant 0 \qquad X_2 \geqslant 0$
Dual:	minimize	$\hat{Z} = 1,000Y_1 + 600Y_2$
	subject to	$2Y_1 + 4Y_2 \geqslant 1$
		$10Y_1 + 5Y_2 \geqslant 4$
	non-negativity	$Y_1 \geqslant 0 \qquad Y_2 \geqslant 0.$

Since our dual programme contains no more than three decision variables, it can be solved using the graphical method exactly in the same way as the solution to the primal problem. The solution to the dual programme will have the following properties which are known as the duality theorem.

Theorem I   The optimal values of the primal and the dual objective functions are always identical, provided that optimal feasible solutions do exist. That is, optimal $\hat{Z}$ = optimal Z.

Theorem II   (a) If a certain decision variable in the primal programme is optimally non-zero then the corresponding slack variable in the dual programme must be optimally zero.

(b) If a certain slack variable in the primal programme is optimally non-zero then the corresponding decision variable in the dual programme must be optimally zero.

ECONOMIC INTERPRETATION OF DUAL SOLUTIONS

The essence of the duality theorem is the valuation or the pricing of resources. In giving an economic interpretation to dual solutions, we must know the nature of dual variables as well as their unit of measurement. Our primal problem is the maximization of the value of national output, and the objective function Z is expressed in money terms. By Theorem I, therefore, the dual objective function

$$\hat{Z} = 1,000Y_1 + 600Y_2$$

must also be in money terms.

But 1,000 and 600 are in physical units denoting resources. Therefore, the coefficients Y_1 and Y_2 in the dual objective function must denote some price variables per unit of each resource. Since at these prices cost is minimized, these imputed or accounting prices can be regarded as efficiency prices. They measure the opportunity cost of using resources. Therefore, $Z = \hat{Z}$ means that, in the optimal solution, the total product must be imputed or allocated in its entirety to the resources via accounting prices. The non-negativity conditions in the dual programme imply that opportunity costs *cannot* be negative. By Theorem II, if there is a slack in the use of a certain resource (i.e. a non-zero solution for a slack variable in the primal programme) then its opportunity cost is zero. This is an argument used by many economists to treat the opportunity cost of labour as zero in labour surplus LDCs.

Let us now consider the dual constraints. The coefficients 2 and 4 in the first constraint describe the amount of each resource (labour and capital) required to produce a unit of the first (agricultural) product. Since Y_1 and Y_2 are the opportunity costs of labour and capital, the first constraint is the total opportunity cost of producing one unit of agricultural product. Since 1 is a coefficient in the primal objective function, it denotes the per unit contribution of the first product to the objective function. (By our assumption, 1 is the per unit price of agricultural product.) Therefore, the first constraint requires that the opportunity cost of producing the agricultural product be imputed at a level at least as large as its net contribution to the national output. Similarly, the second constraint requires that the opportunity cost of producing the manufactured product be imputed at a level at least as large as its net contribution to the national output (which in this case is 4). The implication of this requirement is that if the marginal opportunity cost of producing a product is not equal to its marginal contribution to net returns then the resource allocation is not optimal.

One can easily see that the theoretical foundation of shadow pricing is Pareto optimality. It attempts to approximate prices that would prevail under perfect competition. Notice, however, that the solutions or shadow prices are dependent upon the objective function that has been chosen. What happens, for example, if distribution of income is also an objective of the society? In that case, a single objective programming model cannot allow us to judge the relative significance of the two objectives. Depending on the weights the policy makers give to growth maximization and distribution, the shadow prices will vary.

One of the deficiencies that LP shares with other multi-sectoral models is that resource availability and different parameters of relationships are assumed constant. We have seen in the subsection on the choice of technique how the problems associated with the assumption of a fixed coefficient production function can be dealt with by incorporating activity analysis. However, there still remain at least two problems arising from the assumption of fixed resource supply and other parameters. First, the development plan itself aims at deliberate changes in parameters of relationships and at enhancing the supply of resources. Second, it does not allow for uncertainty. It is common to attribute the failure of development plans to rigidity and their unpreparedness in uncertain events. Therefore, it is often suggested that the planner should check the extent to which the model estimates are sensitive to changes in parameters. This is also necessary because the parameters of relationships may not be known precisely. So, it is advisable to find a range of parameter values for which the optimal model solution remains unchanged. The basic elements in sensitivity analysis are discussed in the appendix to this chapter.

FURTHER READING

Dorfman (1953) gives an early, non-technical explanation of LP, and the book by Dorfman *et al.* (1958) is the authoritative text on the technique. Blitzer *et al.* (1975) give a useful discussion of LP in the context of developing countries and also discuss the consistency of shadow prices derived from LP. Quaddus and Chowdhury (1990) illustrate a procedure to derive weights for multiple objectives in the social preference function. Cohen (1978) has an extensive treatment of multi-objective programming in the planning process.

APPENDIX: THE SIMPLEX METHOD

It was mentioned in the text that to solve LP problems with more than three choice or decision variables we need to use the simplex method. The simplex method is an iterative procedure which makes use of corner points of the basic feasible region. The iteration continues until the optimal solution is obtained. To illustrate the procedure let us use the example provided in the text. That is,

$$\begin{aligned}
\text{maximize} \quad & Z = X_1 + 4X_2 \\
\text{subject to} \quad & 2X_1 + 10X_2 \leqslant 1{,}000 \\
& 4X_1 + 5X_2 \leqslant 600 \\
& X_1 \geqslant 0 \qquad X_2 \geqslant 0.
\end{aligned}$$

The first step in the simplex method is to convert all inequality constraints into the equality form. Remember that a '$\leqslant$' type inequality implies a possibility of slacks in the resource use. Therefore, in the case of these constraints, a slack variable for each constraint is added to the left-hand side of the inequality. The slack variables, however, will have zero coefficients in the objective function. Thus, our LP problem can be written in standard form as

$$\begin{aligned}
\text{maximize} \quad & Z = X_1 + 4X_2 + 0S_1 + 0S_2 \\
\text{subject to} \quad & 2X_1 + 10X_2 + S_1 = 1{,}000 \\
& 4X_1 + 5X_2 + S_2 = 600
\end{aligned}$$

where $S_1 \geqslant 0$ and $S_2 \geqslant 0$ are slack variables.

In the standard form we have four variables (X_1, X_2, S_1, S_2) and two constraint equations. That is, we have only two equations to solve for four unknowns. This is not possible as the maximum number of unknowns that can be solved is equal to the number of independent and consistent equations. So, we can solve for only two variables. We can arbitrarily set values for $n - k = 2$ (number of variables minus the number of equations) variables and solve for the rest. Thus, there will be $4C_2(= nC_{n-k}) = 6$ sets of solutions. To try six sets for the optimal solution is certainly inefficient. At this stage, the simplex method uses two properties of the graphical solution. First, we have seen in the graphical solution that the basic feasible region has four corners each of which represents a basic feasible solution out of which only one is the optimal solution. Therefore, we can say that out of six sets only four are feasible and two are infeasible solutions involving negative values of variables (slack or choice). The simplex method uses this information and investigates only the corner points for optimality and thereby economizes the iteration procedure. Second, for LP problems with '$\leqslant$' inequalities the origin ($X_1 = 0$, $X_2 = 0$) is a basic feasible solution. Therefore, the simplex iteration can start with $X_1 = X_2 = 0$ and find values for $S_1(=1,000)$ and $S_2(=600)$. The variables that are set equal to zero in an iteration are called basic variables.

Let us now examine the simplex iteration to solve our LP problem. For convenience, data are arranged in a table called a simplex tableau. The initial tableau is presented in Table 5.1. It is constructed from the coefficients of variables in standard form. Note the following characteristics of the initial tableau.

1 The variables that do not appear explicitly in an equation are considered to have zero coefficients in that equation. Thus, zeros in the S_2 column of the first row and the S_1 column of the second row indicate that S_2 and S_1 do not appear in the first and second equations respectively.
2 The initial tableau is constructed by setting $X_1 = X_2 = 0$ (the origin in our graphical solution). That is, S_1 and S_2 take the status of basic variables and X_1 and X_2 non-basic variables. If $X_1 = X_2 = 0$, i.e. no production is taking place, then resources will be unutilized and hence the slack variables take the value of constraints. If we set

Table 5.1 The initial simplex tableau

C_j	Basic variables	Right-hand side	C_j			
			1	4	0	0
			X_1	X_2	S_1	S_2
0	S_1	1,000	2	10	1	0
0	S_2	600	4	5	0	1
	Z_j	0	0	0	0	0
	$C_j - Z_j$		1	4	0	0

$X_1 = X_2 = 0$ in the constraints then $S_1 = 1,000$ and $S_2 = 600$ which are recorded under the column 'Right-hand side' (RHS).

3 The coefficients of slack variables (in an LP problem with '$\leqslant$' type inequalities) form an identity matrix.

4 The matrix of coefficients formed by choice or decision variables (X_1, X_2) is called the 'body matrix'. The elements of the body and identity matrices represent marginal rates of substitution between variables in the solution (basic variables) and the variables heading the columns. For example, the element 2 in S_1 row and X_1 column implies that S_1 must be decreased by 2 units if 1 unit of X_1 is added.

5 The Z_j row represents the total value of the objective function (GDP in our example) from the solution. This is zero for the initial tableau (for maximization problems).

6 The $C_j - Z_j$ row represents the net contribution from adding one unit of a variable. For a maximization problem, the presence of at least one positive number in the $C_j - Z_j$ row indicates that the objective function can be improved. Thus, the optimal solution is obtained when there is no positive number in the $C_j - Z_j$ row. For a minimization problem, the presence of at least one negative number in the $C_j - Z_j$ row indicates that the objective function can be improved in a subsequent iteration and the optimal solution is attained when there is no negative number in this row.

Since there are positive numbers in the $C_j - Z_j$ row of the initial tableau, we have to try a second iteration which will improve the solution. That is done by replacing a basic variable in the initial solution with a non-basic variable. Therefore, we have to determine the 'entering' and 'departing' variables.

Entering variable Looking at the C_j row, we can see that the largest per unit contribution to the objective function (GDP) is made by X_2. So, X_2 is the entering or replacing variable.

Departing variable Divide the RHS values by the corresponding entries in the replacing column (heading X_2 in this case). Thus,

$$S_1 \text{ row} \qquad 1,000/10 = 100 \text{ units of } X_2$$
$$S_2 \text{ row} \qquad 600/5 = 120 \text{ units of } X_2.$$

These ratios set the upper limit for the entering variables. For example, given the resources the maximum amount of X_2 that can be produced is 100. This will completely exhaust the first resource (labour) and take only $(5 \times 100 =)$ 500 units of the second resource (capital) that will be released from the production of S_1. On the other hand, if 120 units of X_2 are produced, $(10 \times 120 =)$ 1,200 units of labour will be required which is more than what is available. Therefore, the variable corresponding to the smaller ratio is the departing or replaced variable.

How to recompute the simplex tableau

The X_2 (the replacing) row of the revised tableau is computed by dividing each number in the replaced (S_1) row by the element in the replacing (X_2) column, which is 10 in this example. The entries in the X_2 row are therefore

$$100 \qquad 2/10 \qquad 1 \qquad 1/10 \qquad 0.$$

All the remaining row(s) are computed using the formula

$$\begin{matrix} \text{element in} \\ \text{new row} \end{matrix} = \begin{matrix} \text{element in} \\ \text{old row} \end{matrix} - \left(\begin{matrix} \text{element of old} \\ \text{row in the} \\ \text{replacing column} \end{matrix} \right) \left(\begin{matrix} \text{corresponding} \\ \text{new element in} \\ \text{replacing column} \end{matrix} \right).$$

Therefore, the S_2 row entries in this example are

$$\begin{aligned}
600 - (5)(100) &= 100 \\
4 - (5)(2/10) &= 3 \\
5 - (5)(1) &= 0 \\
0 - (5)(1/10) &= -1/2 \\
1 - (5)(0) &= 1.
\end{aligned}$$

The Z_j row is calculated as follows.

$$\begin{aligned}
\text{total value of GDP} &= 4(100) + 0(100) = 400 \\
X_1 &= 4(2/10) + 0(3) = 4/5 \\
X_2 &= 4(1) + 0(0) = 4 \\
S_1 &= 4(1/10) + 0(-1/2) = 4/10 \\
S_2 &= 4(0) + 0(1) = 0.
\end{aligned}$$

The revised simplex tableau is presented in Table 5.2.

Table 5.2 is not optimal as there is a positive entry in the $C_j - Z_j$ row implying that GDP can be increased further by including X_1. In this case, X_1 replaces S_2 as a basic variable. The revised and optimal tableau is presented in Table 5.3.

Table 5.2 The revised simplex tableau

C_j	Basic variables	Right-hand side	1	4	0	0
			X_1	X_2	S_1	S_2
4	X_2	100	$\dfrac{2}{10}$	1	$\dfrac{1}{10}$	0
0	S_2	100	3	0	$\dfrac{-1}{2}$	1
	Z_j	400	$\dfrac{4}{5}$	4	$\dfrac{4}{10}$	0
	$C_j - Z_j$		$\dfrac{1}{5}$	0	$\dfrac{-4}{10}$	0

Table 5.3 The optimal tableau

C_j	Basic variables	Right-hand side	1	4	0	0
			X_1	X_2	S_1	S_2
1	X_1	33.3	1	0	$\dfrac{-5}{30}$	$\dfrac{1}{30}$
4	X_2	93.34	0	1	$\dfrac{4}{30}$	$\dfrac{-2}{30}$
	Z_j	406.66	1	4	$\dfrac{11}{30}$	$\dfrac{2}{30}$
	$C_j - Z_j$		0	0	$\dfrac{-11}{30}$	$\dfrac{-2}{30}$

The absence of any positive entry in $C_j - Z_j$ indicates that Table 5.3 is the optimal tableau and the optimal solution is

$$X_1 = 33.3 \qquad X_2 = 93.3 \qquad S_1 = 0 \qquad S_2 = 0.$$

The associated GDP $= Z = 406.67$, which is the same as our previous graphical solution.

Note that the body matrix of the initial tableau has become an identity matrix in the final (optimal) tableau and the identity matrix is the inverse of the initial body matrix. Thus, the computational routine for the simplex method is based on matrix algebra and consists essentially of obtaining an inverse matrix in order to solve a set of simultaneous linear equations.

The use of artificial variables

The above simplex procedure is described for problems involving ' $\leqslant$ ' type inequalities. In such cases, we can use the origin (zero values for choice variables) as a starting point for iteration. However, if the inequalities are of a ' $\geqslant$ ' type which frequently occurs in a minimization problem, the origin does not form a basic feasible solution as can be seen in the example below (Figure 5.5):

$$\begin{aligned} \text{minimize} \quad & C = 2X_1 + 10X_2 \\ \text{subject to} \quad & 2X_1 + X_2 > 6 \\ & 5X_1 + 4X_2 > 20. \end{aligned}$$

The basic feasible region is given by the area ABC.

In such cases, the initial solution has to be searched for. This is done in the following way.

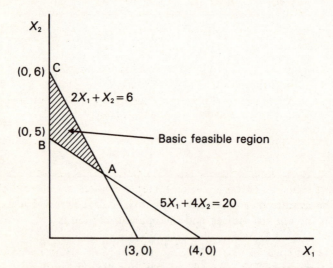

Figure 5.5 Graphical solution to a minimization problem

86

1 A slack variable is subtracted and an artificial variable is added to the left-hand side of the constraint which is of a '$\geqslant$' type for both maximization and minimization problems.
2 The slack variables are given zero coefficients in the objective function for both maximization and minimization problems.
3 The artificial variables are included in the objective function with a positive coefficient for minimization problems and a negative coefficient for maximization problems. These coefficients must be very large in absolute value. This acts as a penalty for including the artificial variables in the final solution.

APPENDIX: SENSITIVITY ANALYSIS

In this appendix we shall illustrate how sensitivity analysis is performed for LP problems. First, we shall consider changes in the parameters of the objective function, and second, changes in resource availability.

Changes in the objective function coefficients

The parameters of the objective function in our example represent the prices of sectoral output. The sensitivity analysis will allow for a range of values for those prices for which the optimal solution will remain unchanged. In terms of the optimal tableau, the RHS values will *not* change, but the C_j row will have a range of values for each column. The procedure to find this range for C_j is as follows.

Step 1   Replace C_j by $C_j + \delta_j$ in the objective function.
Step 2   Find the new Z_j and $C_j - Z_j$ rows.
Step 3   Apply the condition of optimality, i.e. $C_j - Z_j$ must have non-positive entries for maximization.

To illustrate this procedure, we need to refer to the optimal tableau (Table 5.3).

First, let us find the range for C_1, i.e. the coefficient of X_1 (the price of sector 1's output). Therefore step 1 requires us to rewrite the objective function as

$$\text{maximize} \qquad Z = (1 + \delta_1)X_1 + 4X_2 + 0S_1 + 0S_2.$$

The next step is to find the Z_j and $C_j - Z_j$ rows. Using the values from the optimal tableau, we obtain Z_j and $C_j - Z_j$ as follows.

total value of
$$\text{GDP} = Z = (1 + \delta_1)33.3 + 4(93.3) = 406.7 + 33.3\delta_1$$

The Z_j rows are as follows:

$$
\begin{array}{lll}
X_1 & 1(1 + \delta_1) + 0(4) = 1 + \delta_1 \\
X_2 & 0(1 + \delta_1) + 1(4) = 4 \\
S_1 & -5/30(1 + \delta_1) + 4/30(4) = 11/30 - 5/30\delta_1 \\
S_2 & 1/3(1 + \delta_1) - 2/30(4) = 2/30 + 1/3\delta_1 \\
\end{array}
$$

$$C_j - Z_j \quad 0 \quad 0 \quad -11/30 + 5/30(\delta_1) \quad -2/30 - 1/3(\delta_1).$$

The optimality condition requires that $C_j - Z_j$ must have non-positive entries for maximization problems.
Therefore,

$$-11/30 + 5/30\delta_1 \leqslant 0 \qquad (A5.1)$$

and

$$-2/30 - 1/3\delta_1 \leqslant 0 \qquad (A5.2)$$

From (A5.1)

$$\delta_1 \leqslant 2.2$$

and from (A5.2)

$$\delta_1 \geqslant -0.2$$

So,

$$-0.2 \leqslant \delta_1 \leqslant 2.2.$$

Therefore, the limit or range for the price of X_1 is

$$1 - 0.2 \leqslant C_1 \leqslant 1 + 2.2$$
$$0.8 \leqslant C_1 \leqslant 3.2.$$

The solution $X_1 = 33.3$ and $X_2 = 93.3$ remains optimal for the limit of C_1 between 0.8 and 3.2. However, the value of GDP will change by $33.3\delta_1$, i.e.

$$Z = 406.7 + 33.3\delta_1.$$

Depending on the value of δ_1, $-0.2 \leqslant \delta_1 \leqslant 2.2$, the value of GDP will deviate from the original optimal value.

Similarly, we can find the limit for C_2 as

$$1.25 \leqslant C_2 \leqslant 5.$$

If the value of C_2 varies between 1.25 and 5, the optimal solution $X_1 = 33.3$ and $X_2 = 93.3$ will remain unchanged with the value of GDP = $Z = 406.7 + 93.3\delta_2$.

In terms of a graphical solution, this amounts to varying the slope of the objective function (iso-GDP line) within a range of values. But the iso-GDP line will still pass through the same corner point of the basic feasible region yielding the same optimal solution for the decision variables X_1, X_2. This is shown in Figure 5.6. GH is the original iso-GDP line which yielded the optimal solution ($X_1 = 33.3$, $X_2 = 93.3$). The iso-GDP line will rotate at the optimal point E within the limits of the parameters. For the upper limit 3.2 of C_1 (the unit price of X_1), the iso-GDP line will be steeper and coincide with the capital constraint CD. For the lower limit 0.8 it will be flatter and coincide with the labour constraint AB. This, therefore, defines the boundary for rotation of the iso-GDP line without violating the

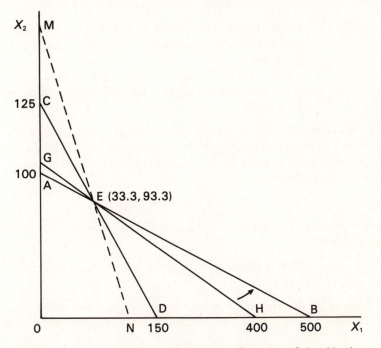

Figure 5.6 Optimal solution and the range of rotation of the objective function

optimality condition (i.e. the iso-GDP line must be tangent with the basic feasible region). If, for example, the iso-GDP line becomes MN whose slope is steeper than that of the capital constraint CD, then a portion of it (EN) will lie inside the basic feasible region. In that case, point E no longer remains optimal as the iso-GDP line will not be tangent with the basic feasible region at E.

Changes in the resource availability: right-hand side of constraints

Sensitivity analysis must also be carried out for changes in the availability of resources. For example, we must know what will happen to our optimal solution if some resources are not available up to the expected level. This often happens in the case of foreign exchange or aid dependent capital projects. In terms of our simplex solution, it amounts to finding a range of RHS values for which the optimality of the current tableau is unaffected. The procedure is as follows.

Step 1 Replace the constraint values by $b_i + \Delta_1$, where b_i is the initial value of the resources (e.g. the initial availability of, say, labour).

Step 2 Find the new RHS values for the optimal tableau.

Step 3 Apply feasibility conditions, i.e. RHS values must be non-negative.

We shall illustrate the procedure by changing the value of the second constraint only. That is, we assume that the availability of resource capital changes by Δ_2. In that case, the second constraint equation becomes

$$4X_1 + 5S_2 + S_2 = 600 + \Delta_2.$$

The next step is to find the RHS in the optimal tableau. The way to do this is

$$b_k^* = \sum_{i=1}^{m} \text{(optimal coefficients of } i\text{th } initial \text{ basic variable in the } k\text{th row) } (b_i)$$

where b_k^* is the RHS value in the optimal tableau. The *initial* basic variables were S_1 and S_2. Their coefficients in the optimal tableau are

X_1 row	$-5/30$	$1/3$
X_2 row	$4/30$	$-2/30$.

Therefore,

$$b_1^*(1,000)\left(\frac{-5}{30}\right) + (600 + \Delta_2)\left(\frac{1}{3}\right)$$

$$= \frac{1,000}{30} + \frac{1}{3}\Delta_2$$

$$b_2^* = (1,000)\left(\frac{4}{30}\right) + (600 + \Delta_2)\left(\frac{-2}{30}\right)$$

$$= \frac{2,800}{30} - \frac{2}{30}\Delta_2.$$

In order for the solution to be feasible, $b_1^* \geqslant 0$ and $b_2^* \geqslant 0$. That is,

$$\frac{1,000}{30} + \frac{1}{3}\Delta_2 \geqslant 0 \qquad (A5.3)$$

and

$$\frac{2,800}{30} - \frac{2}{30}\Delta_2 \geqslant 0. \qquad (A5.4)$$

From (A5.3) we have

$$\Delta_2 \geqslant -100$$

and from (A5.4) we have

$$\Delta_2 \leqslant 1,400.$$

Therefore, $-100 \leqslant \Delta_2 \leqslant 1,400$ which means the availability of resource capital can vary within the range $600 - 100 = 500$ and $600 + 1,400 = 2,000$, given the supply of the other resource, labour.

In terms of Figure 5.6, the capital constraint line can move parallelly to the right up to the point B. Beyond point B, the supply of capital becomes redundant as the maximum amount that can be used is set by the availability of labour. Likewise, the capital constraint line can move parallelly to the left as far as point A. The reduction of capital below this level will make it an absolute constraint and labour supply redundant. Therefore, if the capital constraint line moves parallelly between the points A and B, both the constraints will remain relevant. Similarly, we can find the limit of variations for the supply of labour, keeping the capital constraint fixed.

As one of the constraints is relaxed or tightened, the optimal solution will change. For example, if capital is raised up to its maximum limit, the optimal solution will be at point A, with $X_1 = 500$ and $X_2 = 0$. Accordingly, the value of GDP will be

$$Z^* = 1(500) + 4(0) = 500.$$

If we divide the change in the value of the objective function by the maximum allowable increase in one of the resources, keeping all other constraints fixed, we obtain the shadow price of that resource. The shadow price of capital in our example is therefore

$$\frac{500 - 406.67}{1,400} = 0.07.$$

It is interesting to note that this is the optimal value for the dual variable Y_2. Figure 5.7 depicts the graphical solution to the dual problem:

minimize $\quad Z = 1,000Y_1 + 600Y_2$
subject to $\quad 2Y_1 + 4Y_2 \geqslant 1$
$\qquad\qquad\quad 10Y_1 - 5Y_2 \geqslant 4.$

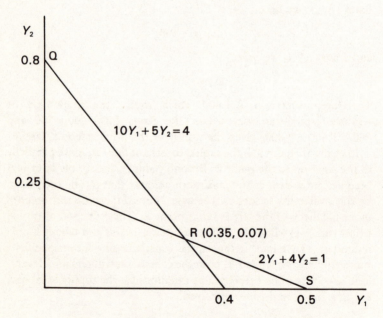

Figure 5.7 Graphical solution to the dual problem

The basic feasible region is given by QRS and the objective function is minimized at R, yielding the optimal solution $Y_1 = 0.36$ and $Y_2 = 0.07$. Recall that dual variables are shadow prices of resources.

Sensitivity analysis can also be carried out for changes in the coefficients defining the constraints (left-hand side of the constraint equations). One important source of changes in constraint coefficients is technological change. In general, sensitivity analysis consists of adding a new variable to the parameter under consideration (RHS values, constraint coefficients and objective function coefficients) for which the optimal solution is not changed.

6

COMPUTABLE GENERAL EQUILIBRIUM MODELS

INTRODUCTION

In the previous chapter, we saw how the input–output model can be extended into the linear programming (LP) framework to obtain the optimal solution from a consistency model. It was noted that the solution to the maximizing LP problem replicates the outcome of the cost minimization/output maximization exercise of profit-maximizing producers in a perfectly competitive production economy. That is, the solution to the LP problem yields, as a by-product, shadow or scarcity prices of resources. Given a set of commodity prices, these shadow prices of resources ensure the efficient allocation of resources as a dual or mirror image of the profit maximization exercise of the perfectly competitive producers. However, factor prices are crucial determinants of factor income and hence influence the structure of demand in an economy where individuals take decisions about their consumption bundle.

The standard LP solution cannot ensure that the structure of demand implied by the dual solution will be consistent with the quantity supplied obtained from the primal solution. If, for example, the demand implied by the dual solution is deficient, the price of output should fall and a new solution for output must be found. But in the standard LP problem, the commodity prices are considered given (exogenous) and there is no mechanism in the solution routine to capture the endogenous interaction of prices and quantities.

This problem would not arise in a centrally planned economy. For in such an economy the structure of demand is not an outcome of an individual's utility-maximizing (with given income) exercise; rather it is the central authority/planners who distribute the output produced and hence any level of factor incomes is consistent with output supplied. Alternatively, in a decentralized socialist economy, optimality in both production and exchange can be obtained if the

94

state owns all the capital and rents it out to bureaucratically managed enterprises whose managers will be instructed to maximize profits of their enterprises. The state, as owner of capital, will receive all the income of each enterprise, net of wages and raw material costs. It should then distribute some of this income to households to supplement their labour incomes or tax the labour incomes in order to ensure that the demand structure is consistent with output supplied.

But developing countries are neither centrally planned nor socialist economies. They are decentralized and mixed economies where individuals (both producers and consumers) interact with each other in order to maximize their profit and utility. What is missing in the standard optimization type planning models is this interaction among individuals as producers and consumers. In other words, there is no feedback between demand and supply. Following Johansen's pioneering work on the economy of Norway, the computable general equilibrium (CGE) models seek to establish the feedback between demand and supply and thereby endogenize both relative prices and quantities. They stress the general equilibrium feedback mechanisms and autonomous decision-making by economic agents.

THE CGE MODEL APPROACH

The essential features of a CGE model are described by Adelman and Robinson (1988: 24–5) as follows.

> A CGE model works by: (1) specifying the various actors in the economy (for example, firms, households, government and the rest of the world); (2) describing their motivation and behaviour (utility maximisation for consumers and profit maximisation for firms); (3) specifying the institutional structure, including the nature of market interactions (competitive markets for goods and labour); and (4) solving for the equilibrium values of all endogenous variables. The model simulates the working of a market economy and solves for a set of prices (including wages, product prices, and perhaps, the exchange rate) that clears all markets (including markets for labour, commodities and foreign exchange).

The third feature above might seem to imply that CGE models deal only with perfectly competitive economies where markets clear instantaneously and governments do not interfere. But they can quite easily handle imperfectly competitive behaviour (e.g. mark-up pricing by the firms and quantity adjustments) and widespread government interventions (e.g. tariffs and quotas on international trade). In fact, it is now increasingly realized that CGE models are much better suited to planning and policy analysis in the mixed economies which characterize most developing countries. Since Adelman and Robinson's (1978) first CGE model for Korea, more than seventy CGE models have been constructed for thirty developing countries to explore issues such as income distribution, IMF stabilization programmes and trade liberalization (Bandara 1991a).

There are two broad categories of CGE models as applied to developing countries. One class of CGE models follows Johansen's work and focuses mainly on the micro-economic aspects of developing countries. Their main concern is real variables such as sectoral growth, employment and investment and structure of production. In short, this class of models concentrates on the resource allocation issues which are central in the medium- to long-run development planning strategy. The specific issues that are analysed in these models include exchange rates, taxes, tariffs and subsidies that affect relative prices and micro-economic incentives and resource allocations. Robinson (1991) calls them 'neoclassical structuralist' CGE models. They are neoclassical in spirit in so far as demand and supply functions are derived from economic agents' constrained optimization behaviour. However, they are 'structuralist' in so far as they replace the neoclassical assumption of market-clearing (through wage–price flexibility) with non-market clearing due to structural rigidities at the micro level. Such rigidities are manifested in low elasticities of substitution between factors, the lack of response of supply to changes in prices (low supply elasticities), underdeveloped markets (e.g. the capital market) and administrative restrictions on crucial variables (e.g. interest rate ceilings, minimum wages).

The second class of CGE models for developing countries are rooted in the Kalecki–Kaldor–Keynes tradition. Robinson (1991) categorizes them as 'macro-structuralist' CGE models as they are driven by macroeconomic relationships such as aggregate savings and investment functions which may not necessarily have micro-economic foundations in constrained optimization behaviour. In other words, unlike microstructuralist models, typically these models do not derive macroeconomic relations from micro functions, by, for example, aggregating individual household demand functions to derive the consumption function. They justify this approach in terms of the empirical regularities of such relationships. Their main concern is consequences of macro-economic disequilibria, e.g. what happens to aggregate and sectoral output and employment if government expenditure exceeds revenue or if the current account of balance of payments is in deficit. This class of models is macro in nature, but structuralist in the sense that the models allow nominal variables (such as budget deficits and current account deficits) to affect real variables (such as sectoral output and employment). In other words, they reject the neoclassical assumption of neutrality between nominal and real sectors. In the neoclassical model, all that matters is relative prices. As long as relative prices remain unaffected, any change in the absolute price level cannot affect equilibrium in the real sector. The implication is that output is always at full-employment level. In contrast, macro-structuralist modellers believe that relative prices in developing countries do not work according to the neoclassical paradigm. In macro-structuralist models, equilibrium is achieved through quantity adjustments rather than by changes in relative prices. The reasons for quantity adjustment lie in the existence of factors such as monopolistic industrial structure and mark-up pricing, segmented capital markets (formal and informal), financial repression (interest rate ceilings) and fixed exchange rates.

The distinction between these two types of CGE models can be conceptualized to some extent by referring to the LP- and SAM-based models discussed in previous chapters. As we saw, LP is a tool for examining policy impacts on sectoral output levels. Its focus is on efficient resource allocation and relative (shadow) prices. On the other hand, one of the main concerns of SAM models is to analyse policy impacts on the savings–investment balance via income distribution effects.

However, the distinction between macro- and micro-economic issues is somewhat arbitrary. For example, sectoral outcomes in LP can be aggregated to obtain aggregate (macro) magnitudes, while SAM-based models also derive the sectoral implications of the alternative macro scenario. This arbitrariness can be explained by using the example of the oil price shocks of the 1970s. As is well known, the two oil price shocks created serious macro-economic imbalances in most oil importing developing countries. These imbalances were manifested in large current account deficits which put pressure on the policy makers to devalue the exchange rate. In an attempt to defend their currencies (most developing countries follow a fixed exchange rate system), countries ran down their foreign reserves. The defence of the exchange rate was also motivated by a desire to prevent imported inflation. However, this created a policy dilemma. The decline in foreign reserves would have reduced the money supply with a contractionary impact on output and employment. In order to avoid such consequences, most governments pursued an expansionary fiscal policy. As a result, the government budget deficit became very large and inflation soared. The oil price increases also had micro-economic effects. For example, both final and intermediate users of oil substituted other cheaper energy for oil. At the industry level, this meant a change in the technology of production and the production mix. Depending on the substitution possibilities, one can also expect a rise in the relative price of oil-based products. This rise in price can be regarded as imported inflation.

But these macro- and micro-economic effects are not mutually exclusive or independent of one another. For example, the over-valuation of exchange rates affected the relative price of tradables and non-tradables sectors and biased the resource allocation in favour of non-tradables. This, in turn, made the balance-of-payments adjustments more difficult. Similarly, the growth of money supply due to the budget deficit accommodated the rise in the price of oil-based products without a corresponding decline in other prices, thus adding to inflation.

The oil price crisis, therefore, makes the interdependence of structural (micro) and stabilization (macro) issues clear. In practice, medium-term planning or structural strategies are often formulated to include some macro objectives such as a reduction in the inflation rate or current account deficit, and also depend crucially on such aggregate variables as the saving rate and resource flows, which are normally classified as macro variables. It is therefore necessary to formulate models that allow one to analyse this interdependence between the short-term stabilization (macro) and medium-term structural/planning (micro) issues in order to explore questions of policy trade-offs and policy effectiveness. However, the inclusion of nominal flows in the CGE models is not a straightforward matter and the attempts to integrate short-term macro models with multi-sectoral CGE models are somewhat *ad hoc* in nature (Robinson and Tyson 1984: 89). But some more recent work appears more promising and a group of CGE models, which Robinson (1991) calls 'financial' CGE models, are emerging which have features of both 'neoclassical' and 'macro'-structuralist CGE models.

AN ILLUSTRATIVE CGE MODEL

In this section we shall illustrate the basic features of a Johansen type CGE model. For simplicity we shall consider a two-sector, two-factor closed economy and will not make any distinction between commodities and industries. Nor will we make a distinction between different types of final uses, but instead we consider only final consumption.

A stylized Johansen type CGE model typically involves the development of

1 a series of equations representing household and other types of final demands for commodities;
2 a series of demand equations for intermediate and primary factor inputs;
3 a series of pricing equations relating commodity prices to costs;
4 a series of market-clearing equations for primary factors and commodities;
5 income formation equations to translate factor incomes into household incomes.

Final demands

In the simple input–output model, final demands are assumed to be exogenously or administratively determined. In the SAM-based multisectoral model final demands are endogenized by assuming that sectoral final demand deliveries are in fixed proportions to total final demand. The total final demand (e.g. aggregate consumption) is normally modelled as a function of disposable income in line with short-term macro-economic models, rather than being derived from economic agents' optimization behaviour.

On the other hand, as mentioned earlier, in the neoclassical CGE models final demands are derived from consumers' utility maximization behaviour. We assume that households maximize a Cobb–Douglas type utility function

$$U = X_{1f}^{e} X_{2f}^{1-e}$$

subject to the budget constraint

$$P_1 X_{1f} + P_2 X_{2f} = Y$$

where X_{if} refers to consumption of the ith commodity and P_i to commodity prices. Y is the nominal income. The households' constrained optimizing behaviour yields household commodity demand functions as

$$X_{1f} = \frac{1}{e}\frac{Y}{P_1} \qquad \text{household final demand for commodity 1} \quad (6.1)$$

and

$$X_{2f} = \frac{1}{1-e}\frac{Y}{P_2} \qquad \text{household final demand for commodity 2.} \quad (6.2)$$

Input demands

There are two kinds of inputs – intermediate and primary. They are derived from firms' constrained cost minimization/profit maximization behaviour. If we assume that one unit of value added Y_j is needed to produce one unit of gross output X_j then we can write the production function for gross output as

$$X_j = X_j(X_{ij}, Y_j)$$

where X_{ij} is the intermediate input.

The value-added production function is assumed to display constant returns to scale and is of a Cobb–Douglas type as

$$Y_j = K_j^b L_j^{1-b}$$

where K_j and L_j are the amount of capital and labour employed in industry j.

The total cost will include both intermediate inputs and primary factor costs and can be expressed as

$$C_j = P_1 X_{1j} + P_2 X_{2j} + wL_j + rK_j$$

where w and r are prices of labour and capital respectively.

To produce a given level of output, firms must choose the level of input use in such a way that the total cost is minimized. At this stage, we assume that there is no possibility of substitution between intermediate and primary inputs. This is not an unrealistic assumption given that they perform completely different tasks and it allows us to make the decision to employ primary factors and to use intermediate inputs separable.

We assume that intermediate inputs are combined in fixed proportions to industry output. (We made the same assumption when discussing the input–output and SAM models.) This assumption implies constant returns to scale and does not allow for substitutions between inputs. The intermediate input demand functions, thus, can be written as

$X_{ij} = a_{ij}X_j$ output of industry i used as inputs in industry j

where $a_{ij} = X_{ij}/X_j$ is the amount of intermediate inputs used to produce one unit of output in industry j. This is exactly the same expression that we had in the input–output and SAM-based multi-sectoral models. We can write the intermediate input demand functions more explicitly as

$$X_{11} = a_{11}X_1 \qquad (6.3)$$
$$X_{12} = a_{12}X_2 \qquad (6.4)$$
$$X_{21} = a_{21}X_1 \qquad (6.5)$$
$$X_{22} = a_{22}X_2. \qquad (6.6)$$

However, primary factor demands will be treated differently from the input–output and SAM-based multi-sectoral models. Instead of assuming that primary factors are used in fixed proportions, they will be made functions of relative factor prices and hence there will be substitution possibilities between them.

Since the primary factor choice is independent of firms' intermediate input demands, they will essentially minimize the primary factor cost component $wL_j + rK_j$ of total cost while employing primary factors. The cost minimization/profit maximization behaviour of firms will yield primary factor demands as functions of relative factor prices and output level. The primary factor demands by industry j can be written as

$$L_j = \left\{ \frac{[(1-b)/b]r}{w} \right\}^b X_j \qquad \text{demand for labour}$$

$$K_j = \left\{ \frac{[b/(1-b)]w}{r} \right\}^{1-b} X_j \qquad \text{demand for capital.}$$

Written more explicitly, they become as follows:

$$L_1 = \left\{ \frac{[(1-b)/b]r}{w} \right\}^b X_1 \qquad (6.7)$$

$$L_2 = \left\{ \frac{[(1 - b)/b]r}{w} \right\}^b X_2 \qquad (6.8)$$

$$K_1 = \left\{ \frac{[b/(1 - b)]w}{r} \right\}^{1-b} X_1 \qquad (6.9)$$

$$K_2 = \left\{ \frac{[b/(1 - b)]w}{r} \right\}^{1-b} X_2. \qquad (6.10)$$

Price equations

The above factor demand equations (6.7)–(6.10) can be expressed in per unit form as

$$l_1 = L_1/X_1 = \left\{ \frac{[(1 - b)/b]r}{w} \right\}^b \qquad (6.11)$$

$$l_2 = L_2/X_2 = \left\{ \frac{[(1 - b)/b]r}{w} \right\}^b \qquad (6.12)$$

$$k_1 = K_1/X_1 = \left\{ \frac{[b/(1 - b)]w}{r} \right\}^{1-b} \qquad (6.13)$$

$$k_2 = K_2/X_2 = \left\{ \frac{[b/(1 - b)]w}{r} \right\}^{1-b} \qquad (6.14)$$

These equations give us the demand for labour and capital by industry j to produce one unit of output, and they are functions of relative factor prices.

To obtain unit cost of production we divide total costs by total output. Therefore, the unit or average cost is

$$P_1 = P_1 a_{11} + P_2 a_{21} + w l_1 + r k_1 \qquad (6.15)$$
$$P_2 = P_1 a_{12} + P_2 a_{22} + w l_2 + r k_2. \qquad (6.16)$$

Since the profit maximization/cost minimization hypothesis and constant returns to scale imply that marginal cost equals average cost equals price, the above equations represent the price of one unit of output j. This is once again similar to the cost-determined price equations of the input–output and SAM-based multi-sectoral models. The only difference in this case is that per unit primary factor demands (l_j, k_j) are functions of relative factor prices instead of being constants.

Market-clearing equations

These equations balance demand with supply. The demand for commodities includes final demand as well as intermediate demands. Therefore, we can write the balance equations for commodities as

$$X_1 = a_{11}X_1 + a_{12}X_2 + X_{1f} \tag{6.17}$$

$$X_2 = a_{21}X_1 + a_{22}X_2 + X_{2f}. \tag{6.18}$$

The balance equations for primary factors are

$$L_1 + L_2 = L \tag{6.19}$$

$$K_1 + K_2 = K. \tag{6.20}$$

Income formation

This is the final equation in the model and translates factor income into institutional income. Since in this one-household model the sole household supplies both labour and capital services, the factor income will be the same as household income. Therefore, the household income is

$$Y = wL + rK = \text{GDP}. \tag{6.21}$$

Solution of the model

The solution of a model entails finding the equilibrium values of variables. There are two special features of the solution of the model developed above. First, as mentioned earlier, the neoclassical models assume neutrality between the real and nominal sectors. Thus, the solution of the above model can establish only relative prices and not the absolute price level. As only relative prices matter, we must specify 'relative to what'. That is, we have to choose a 'numeraire' and the solution of the real variables must remain unaffected by the choice of a numeraire. This means that any one price can be chosen as a numeraire (and all other prices expressed in relation to it) as long as they all vary in the same proportion as the numeraire. For example, if the numeraire price is doubled, all other prices must also be doubled so that relative prices remain unchanged. This will leave the solutions for real variables (e.g. equilibrium amount demanded and supplied) unchanged. Choosing one price as a numeraire is known as the 'normalization' procedure.

Second, the solution of a model requires that the number of independent and consistent equations must be equal to the number of variables. However, such models are not interesting from a policy point of view as everything is interdependent, whereas policies are generally determined by administrative decisions outside the model. In the above model, we have twenty-three variables, but only twenty-one equations. Thus, we have to regard two of the variables as exogenous. This is known as 'closure' and solutions will differ depending on the variables chosen to be exogenous or according to the closure rules. Usually the supply of labour and capital are assumed exogenous. However, the choice of closure should be guided by the features of an economy. For example, in most developing countries the interest rate is administratively determined and kept below the market equilibrium level. The existence of surplus labour in most developing countries also implies that the wage rate is not affected by demand for labour and can be regarded as fixed. Therefore, we can treat w and r as exogenous for the solution of the model.

The model will have solutions for each specific value assumed for the exogenous variables. These are known as simulations. Therefore, we can simulate the model for the values of sectoral outputs X_i, sectoral employment L_i, household income Y and consumption X_{if}, and so on, for different values of the wage rate and rental for capital.

There is one more complication in the solution procedure. The model is non-linear in the sense that some variables do not appear in additive form and have a power other than unity. This problem is resolved by expressing the equations in linear percentage forms. An additional advantage of this procedure is that most parameters can be obtained from either an input–output table or a SAM. This implies that a country for which a CGE model is to be implemented must have a reliable and updated input–output table or a SAM.

This illustrative model is admittedly very simple. However, it can be extended for an open economy and to include various structural characteristics of a developing country. We can extend the basic model to incorporate the open economy case in the following way. For simplicity, we assume that all goods are tradables. We further assume that the economy is small and hence the world price ratio is given. The economy can be depicted as shown in Figure 6.1.

The world price ratio is given by WW which is steeper than the domestic price ratio DD. The world price ratio is tangent to the production possibility frontier at A and to the community indifference curve at B. This allows the economy to produce at A and exchange

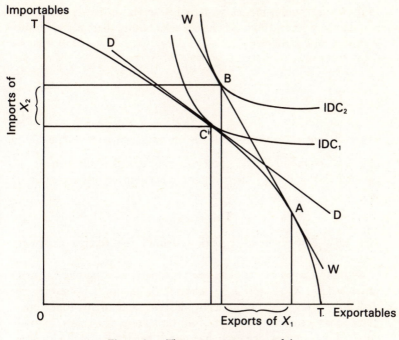

Figure 6.1   The open economy model

X_1 in the world market to obtain X_2 and thereby attain a higher level of utility than in the case of autarky.

The above features of a small open economy do not affect the utility and profit functions of consumers and producers in any essential way. Since all goods are tradables, there is no distinction between imported and domestic goods and hence the substitution possibilities between X_1 and X_2 remain unaffected (regardless of their sources) by the openness of the economy. In other words, the slopes of the community indifference and transformation curves, and hence the final and input demand functions (equations (6.1)–(6.14)), remain unchanged. Therefore, the only equations of the closed economy model that will be affected are the commodity market-clearing equations (6.17)–(6.18). They should be replaced by

$$X_1 = a_{11}X_1 + a_{12}X_2 + X_{1f} + X_e$$
$$X_2 = a_{21}X_1 + a_{21}X_2 + X_{2f} - X_m$$

where X_e is export and X_m is import.

In the open economy model, we have two sets of prices – world prices and domestic prices. Both sets bear a relationship determined by the exchange rate as follows:

$$P_1 = qP_{w1} \tag{6.22}$$

and

$$P_2 = qP_{w2} \tag{6.23}$$

where P_{wi} are world prices and q is the exchange rate.

One final equation in the open economy model is the balance of payments which defines the relationship between exports and imports as

$$P_{w1}X_e - P_{w2}X_m = 0. \tag{6.24}$$

The balance-of-payments equation defines the foreign exchange constraint for an economy.

With the above features of an open economy, we introduce five new variables (X_e, X_m, P_{w1}, P_{w2}, q) and three equations. Therefore, we shall have altogether twenty-eight variables and twenty-four equations. This means we have to treat four variables as exogenous. In most developing countries, the exchange rate is fixed. The small economy assumption also implies that world prices are given. We can assume that exports are determined by world income and hence exogenous for the economy under consideration. These features imply that q, P_{w1}, P_{w2} and X are exogenous.

The assumption of fixed world prices and fixed exchange rates changes the open economy model in one more fundamental way. When this assumption is combined with the assumption of constant returns to scale, equations (6.22) and (6.23) imply that the exogenously determined world prices determine the domestic prices. Once the domestic prices are determined, we obtain the factor prices from the unit price equations. Thus, the fixed world prices make the open economy model a fixed-price rather than a flex-price model. With fixed prices, the market clearing is brought about by quantity adjustments. The quantity adjustment is believed to be an important feature of most developing countries where many prices are sluggish or administratively fixed.

USES OF CGE MODELS

In contrast to the input–output and other multi-sectoral models which capture only quantity relationships (and hence appear to be more suitable for centrally planned economies), the CGE models are able to incorporate market mechanisms and policy instruments that work through price incentives. Since most developing countries operate within the structure of a mixed economy, the CGE models are becoming fashionable among planners. The CGE models are increasingly being used to analyse a wide range of issues. These include tax policy (e.g. direct versus consumption tax), the choice of trade strategy (e.g. import substitution versus export-oriented industrialization), income distribution, structural adjustment to external shocks and long-term growth and structural changes in developing countries. The World Bank's and IMF's recent recommendations relating to trade liberalization and to the introduction of value-added tax are examples of policy prescriptions which have been developed using the results of CGE modelling (see Dervis *et al.* 1982; Mansur and Khondker 1991).

It was suggested earlier that a common analytical framework is needed to harmonize the programmes of the planning and budgetary agencies. This need for harmonization arises as the concerns of these two agencies differ. While the planning agency (the planning ministry) is concerned with long- to medium-term issues relating to such real variables as growth and production structure, the budgetary department is concerned with short-term stabilization. The budgetary agency's (the finance ministry's) main focus is the balance between such nominal flows as balance of payments and public income and expenditure. In short, the responsibility of the planning ministry is supply management and that of the finance ministry is demand management. Therefore, the CGE models (with micro–macro sub-models) can be used to coordinate short-term stabilization and long-term structural adjustment programmes.

In any economy, both supply and demand interact to produce an economic outcome, regardless of their responsiveness to changes in relative prices. In the earlier days of planning, too much attention was given to supply-side issues, on the assumption that developing countries were plagued with all sorts of structural bottlenecks (e.g. a low level of savings, zero price elasticity of supply), such that the 'Keynesian' demand multiplier would work only in nominal terms (Rao 1952). The emphasis on supply considerations was to the almost complete neglect of demand-side issues. In many cases this resulted in serious macro-economic imbalances. For example, attempts to mobilize resources through forced savings led to serious inflation and current account deficits. The balance-of-payments problem was compounded by overvalued exchange rate policy designed to complement the import substitution industrialization programme and to insulate domestic prices from imported inflation.

The IMF's stabilization and structural adjustment programmes since the mid-1970s have been aimed precisely at these problems. One of the main instruments of an IMF programme is devaluation, designed to correct the balance-of-payments problem and to induce structural adjustment. Thus, devaluation is expected to have both nominal and real effects. It raises the domestic price of tradables and thereby induces expenditure switching away from tradables to non-tradables. This also induces resource switching from non-tradables to tradables. Both expenditure switching (nominal flows) and resource switching (real flows) are expected to improve the balance of payments. Since the CGE models (especially the 'macro'-structuralist CGE) include both real and nominal flows, they are well equipped to analyse the impact of IMF-type stabilization programmes.

FURTHER READING

Bandara (1991a) gives a comprehensive survey of the uses of CGE models in development policy analysis, and Bandara (1991b) contains an excellent exposition of CGE models using hypothetical data. Dinwiddly and Teal (1988) provide clear, step-by-step illustrations of CGE modelling and also provide a computer program for solutions. Dervis *et al.* (1982) discuss various aspects of the micro-based approach, and Taylor (1990) illustrates the use of the macro-structuralist approach. Both contain examples of CGE models for specific countries. The theoretical issues pertaining to micro–macro integration are discussed in Robinson and Tyson (1984).

7

COST–BENEFIT ANALYSIS

INTRODUCTION

Cost–benefit analysis (CBA) is perhaps the best known technique for public policy analysis and is widely used by policy makers in both advanced and developing economies. In contrast to the models and techniques discussed so far, which have dealt with macro, sectoral and industry level planning decisions, CBA is applied to the most dis-aggregated level of activity, the individual investment project. As we emphasized at the beginning of the book, the basic problem of planning is that of allocating limited resources among alternative uses in a way that will make the net benefit to the economy as large as possible (benefit, of course, can be defined in a number of ways, depending on the objective function selected). Since resources are limited, choices have to be made, and CBA, or what is often called project appraisal, is a method of evaluating the alternative options in a consistent and comprehensive manner. The basic principle of CBA can be stated simply: if the benefits of a project exceed its costs, where both benefits and costs are measured in the same terms, then the project is worth proceeding with, but if costs exceed benefits the project is rejected.

There are three main elements in any project appraisal. The first involves the *identification* of the relevant benefits and costs of the project. Here, we need to identify the increase in outputs (benefits) which are the result of the project, and the reduction in the supply of inputs (costs) available to the rest of the economy which can be attributed to the project's existence. The benefits and costs are identi-fied, therefore, as the differences in availability of outputs and inputs with and without the project. The second stage involves the *valuation* of the identified costs and benefits in terms of a common yardstick, so that the various outputs and inputs can be added together to give an aggregate valuation of total benefits and total costs. Finally, we need to allow for the fact that the benefits and costs associated with the project are spread over the period of time from when the initial

109

investment is made until the project reaches the end of its productive life. Net benefits that occur some time in the future are less valuable than those that accrue in the present, and the procedure of *discounting* needs to be used to convert net benefits accruing in different time periods into equivalent values that can be added together.

These three processes will be applied when considering any investment decision, whether it is in the private or public sector. There is an important difference, however, in the way in which project appraisal is undertaken in the two sectors. In the private sector the objective will be to maximize profits, and benefits and costs will be identified and valued in terms of the financial receipts and expenditures linked to the project. In the public sector, however, the objective function will be expressed in terms of maximizing net benefits to the economy as a whole, which, as we saw in earlier chapters, could simply mean maximizing the economy's output by making full use of the nation's resources, or it could also include a distributional objective. This means that there will normally be differences between project appraisals undertaken in the public and private sectors. First, the costs and benefits identified with the project will be different: there will often be items included (or excluded) in the public appraisal which are excluded (or included) in the private appraisal. Second, the valuation of the identified costs and benefits will differ, with market prices used in the private appraisals and economic, or shadow, prices used in the public project appraisal. Finally, the discount rate applied to the time-stream of benefits and costs will differ for private and public project appraisal.

Unfortunately, there is no generally accepted terminology to distinguish between the private and public sector approaches to project appraisal. Here we shall use the terms financial project appraisal and economic project appraisal to distinguish between the two main approaches to CBA.

FINANCIAL PROJECT APPRAISAL

In financial appraisal, the tasks of identifying and valuing the project's benefits and costs collapse into one, since the benefits are given by the revenue receipts from the sale of the project outputs and the inputs are given by the costs (expenditures) of production. Market prices are therefore used as the unit of valuation.

The first step in the financial appraisal is to calculate the project's cash flow. This is done by recording on an annual basis the revenues and expenditures for the entire life of the project. The difference between the yearly receipts and expenditures is the net cash flow. Costs are divided into capital and operational costs. Capital costs are those costs incurred in establishing the project, and will include costs of equipment, buildings and land. Operating costs are those incurred in running and maintaining the project, and will include raw materials, labour, utilities and repair and maintenance. The financial benefits to be included in the cash flow are identified and valued as the project's output which is sold at the market price. In all cases, the cost or benefit item is entered into the cash flow at the time it occurs.

The net cash flows consist of forecasted revenues and expenditures over the life of the project. It is now necessary to recognize that cash received in the future is less valuable than cash received immediately. The reason for this is simply that money received in the future rather than the present represents an opportunity cost, in terms of the income that could have been earned by investing the funds in an interest-bearing account or revenue-earning productive activity. This is why borrowers have to compensate lenders for the income they are forgoing, by paying a rate of interest. The rate of interest therefore reflects peoples' preference for money in the future; i.e. it represents individuals' 'rate of time preference'.

In order to combine each year's net cash flow into a single aggregate figure, we need to convert them into equivalent terms. This is done by the process of discounting, which converts future values into an equivalent present period value. We can explain this important process by using a simple example. Suppose a firm or individual is asked to choose between £100 today and £100 next year. The choice will be in favour of £100 today, which can then be placed in a savings account, earning, say, 10 per cent a year. After one year the interest payment will have increased the savings account balance to £110. So the prospect of £100 a year from now is equivalent to only £100 divided by 1.1 = £90.9 in present period terms. This process of reducing future values to their present period equivalent value is called discounting. If the example is extended to a second year, then we need to allow for the fact that interest will be earned on the previous period's interest, increasing the savings balance to £121 (£110 + £11). The payment of £100 two years hence would therefore

be discounted to give a present value of £100 divided by 1.21 = £82.6.
A general expression for calculating the net present value (NPV) is

$$NPV = \sum_{t=0}^{n} \frac{B_t - C_t}{(1 + i)^t} .$$

where B_t and C_t are the benefits (revenues) and costs (expenditures)
in each year t, i is the discount rate (rate of interest) and n is the life
of the project. The calculation of NPV can be easily undertaken using
discount tables.

The technique of discounting allows the analyst to take into account
the differences in the timing of cash flow and thus to assess the
viability of projects with different streams of benefit and costs. Two
methods are commonly used in making this assessment – the NPV
criterion and the internal rate of return criterion. The NPV criterion
follows directly from what has already been discussed, namely that a
project is worth proceeding with if its NPV is positive.

The internal rate of return criterion is less obvious, but in most cases
will give the same decision as the NPV rule. By investing in the
project, the owner of the capital has given up other investment oppor-
tunities. The project must therefore yield a rate of return which is at
least equal to the opportunity cost of the investment funds. Now
suppose the NPV of the project is zero. This means that the project's
returns are just sufficient to repay the principal invested in the project
and the interest payments made to lenders. In this case, when
NPV = 0, the discount rate i is known as the internal rate of return.
Now, recall that the discount rate is the same as the rate of interest
and also that the NPV and discount rate will vary inversely (the higher
the discount rate, the lower the NPV).

The internal rate of return (IRR) rule says that a project is worth
undertaking if the internal rate of return is greater than the cost of
capital (IRR $> i$). But it should be clear that this gives the same decision as the NPV > 0 rule: if NPV > 0, then IRR $> i$, whereas if
NPV < 0, IRR $< i$. This relationship is shown in Figure 7.1.

The final consideration in financial project appraisal concerns
making comparisons between projects. Suppose that we have a port-
folio of investment projects, all with positive NPV, but we are unable
to undertake them all because of limited investment funds. Which
project should be selected? The answer is to select the set of projects
that will give the highest aggregate NPV for the investment funds
available.

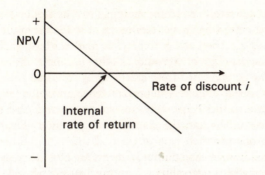

Figure 7.1 The NPV curve with different discount rates

ECONOMIC PROJECT APPRAISAL

In Chapter 5 we saw how the programming method provides a simultaneous solution to the optimum allocation and valuation of resources. This follows from the fact that cost minimization is the mirror image of output maximization. The LP solution yields a set of efficiency prices, or shadow prices, which are equal to the marginal opportunity cost of each output and input.

Exactly the same 'optimal' solution can be obtained from welfare economics, which shows that a perfectly competitive market economy will generate a welfare-maximizing use of the nation's resources, where market prices of outputs and inputs are everywhere equal to marginal opportunity cost. Therefore, if markets were perfectly competitive, they would produce an economic-efficiency outcome, and financial and economic project appraisal would be identical. But we know that perfectly competitive markets seldom, if ever, exist in the real world. Hence the need to undertake an economic appraisal using shadow prices instead of market prices to value the project's benefits and costs.

There are two main sets of factors which cause market prices to diverge from shadow (efficiency) prices. The first is market imperfections or 'failures', and the second is market 'distortions'. Market failures relate to situations where markets for particular goods and services fail to meet the conditions of perfect competition. There are two main categories of market failure where shadow prices will be required – imperfect competition and externalities. Most markets are characterized by a degree of imperfect competition, which means that the market price will exceed the marginal cost of production. Natural monopolies occur when major economies of scale exist in the provision of the goods or services and prevent a free market which would provide an economically efficient outcome. In other instances, the competitive market may be prevented from emerging by collusion among a small number of major producers to prevent other firms from entering the market. Externalities occur when an economic activity has an impact on someone other than the consumer or producer. Environmental damage is one common type of external cost. On the benefit side, training and human capital improvements are often cited as examples where the market undervalues the output being produced. A project may also give rise to external effects which occur through price changes. Those pecuniary externalities may also need to be included in the economic appraisal. Forward linkage effects can occur in the industries that use the project's output and backward linkage effects in industries that supply its inputs if these industries are stimulated by increased demand and higher prices for their outputs or lower prices for their inputs. On the other hand, the project may cause other competing producers to lose market share, and other users of inputs to pay higher prices.

Market distortions relate to situations where market values for particular goods or services diverge from their efficiency values as a result of government intervention in the market in question. Government policies affecting foreign trade are frequently the cause of distortions. For example, the use of tariffs and other non-tariff controls on imports tend to increase the domestic market price of imports and import-substitutes above their 'border' (i.e. cost, insurance and freight (cif)) price. This increase in domestic prices relative to world prices in turn often leads to the maintenance of an overvalued exchange rate, another form of market distortion. Policies directed towards domestic factors of production or outputs can also create 'distorted' (i.e. non-efficient) market prices. For example, minimum wage legislation may result in a market wage rate which is greater than the efficiency (marginal opportunity cost) price of labour. Similarly, controls on interest rates, and on the prices of outputs produced by public enterprises, can cause the market prices to diverge from their efficiency values.

So far we have identified the various factors which make market values inappropriate measures of efficiency values, and which therefore necessitate the use of shadow prices in undertaking economic project appraisal. How then are the shadow prices, which will be substituted for market prices in the economic appraisal, to be calculated?

SHADOW PRICING METHODOLOGY

Several competing methodologies for calculating shadow prices were developed in the late 1960s and early 1970s by various development agencies. The Little and Mirrlees approach (1974) was developed for the Organisation for Economic Co-operation and Development (OECD) and was subsequently extended by the World Bank (Squire and van der Tak 1975) and the UK Overseas Development Administration (1988). An alternative approach was adopted by Dasgupta *et al.* (1972) in developing the methodology used by the United Nations Industrial Development Organisation (UNIDO). Both approaches are very similar in spirit, and in most circumstances yield identical results. We shall not therefore give a detailed comparison of the methods, but will instead concentrate on the Little Mirrlees–World Bank approach which has become the better known and more widely adopted methodology for shadow price estimation.

The first step is to divide the inputs and outputs of a project into traded and non-traded goods. Traded goods are items that enter into international trade-exportables and importables. Non-traded goods are those items which are not traded internationally (goods that could be traded internationally, but which for some reason – e.g. an export ban – are not traded, are treated as non-traded).

The shadow price of a traded good is based on the world market price. The theoretical reasoning behind this is that world prices represent the net benefits associated with a marginal change in production or consumption of the traded good. If a good is traded at a fixed externally set price, then an increase in production will have no effect on the price, domestic demand will not alter and the increase in output will either be exported or substituted for imports. In either case, the effect on the economy as a whole is the foreign exchange earnings or savings.

This approach has the major advantage of allowing the project analyst to calculate shadow prices without having to consider the magnitude of the domestic market imperfections or distortions. World market prices are free of domestic market distortions and therefore give a direct measure of the shadow (efficiency) value of the good.

Since traded goods are valued at world prices, non-traded goods and services have to be valued in the same terms. For non-traded inputs this is done by first estimating the marginal costs of production. These project input costs are then converted to world price values by tracing the direct and indirect foreign exchange costs in the inputs, i.e. by breaking the inputs down into their traded and non-traded inputs, round after round. When this process of decomposition is completed, the analyst is left with traded inputs and the non-traded primary inputs of labour and land. The process of shadow price valuation is then completed by the conversion of non-traded labour and land into world price (foreign exchange) values. For non-traded outputs, the procedure for converting to shadow prices is to determine the traded goods for which they substitute in domestic production. The world prices of these goods can then be used to derive a shadow price value for the non-traded goods output.

These procedures for dealing with non-traded goods are cumbersome and time consuming, and in practice various approximation methods are used in carrying out the project appraisal. These will be explained shortly, but at this point we can note that the Little Mirrlees–World Bank approach is best suited to projects which produce traded outputs and use mainly traded inputs. Where there are many non-traded goods to be considered the approach loses much of its theoretical vigour and practical simplicity.

The calculation of shadow prices requires a great deal of information on market and world prices, as well as knowledge of the economy's macro-economic relations and its micro-economic behaviour. It will often be appropriate for the task to be undertaken by a central planning agency which would advise the individual ministries and agencies undertaking the project appraisals on the shadow price values to be used.

Traded goods

The shadow prices for traded goods are based on their world price (cif for imports and fob for exports). These prices can be expressed in foreign exchange terms, or converted to domestic currency values using the official exchange rate. The world (border) price is then adjusted for the costs of internal transport and distribution that are saved or incurred, with these internal costs expressed in shadow price terms (since transport and distribution services are non-traded goods, they are converted to shadow prices using the standard procedure for non-traded inputs).

The estimation of world prices is not always straightforward. Goods are seldom homogeneous, and there may be differences between the domestically produced and traded goods. The same traded good may sell at different prices in different markets or the price may vary with the volume of sales. A similar import good may be obtainable from different sources at different prices. All of this makes classification and calculation of world prices difficult and introduces an element of judgement into the calculations.

Non-traded goods

For many of the non-traded items, the shadow price is calculated using 'conversion factors'. A conversion factor is the ratio between the market price and shadow price of the good or service concerned, and so if we multiply the market price by the appropriate conversion we obtain an estimate of the shadow price. If these conversion factors are supplied to the project analyst by the central planning agency then the process of decomposition described earlier does not need to be undertaken by each project analyst.

In situations where information on cost composition is not available, or where the non-traded items are a relatively small element in the project appraisal, a general conversion factor is applied to all non-traded inputs and outputs (other than labour, for which a specific conversion factor is almost always estimated). This general conversion is known as the standard conversion factor (SCF). It allows for the general distortion between border (world) prices and domestic prices caused by tariffs, taxes and subsidies, and is estimated from a representative sample of traded goods for the whole economy. The most general formula for the SCF is

$$\text{SCF} = \frac{M + E}{M(1 + t_m) + E(1 + t_e)}$$

where M, E, t_m and t_e, represent total imports (cif), total exports (fob), the average tariff rate and the average export subsidy respectively.

When information is available it is preferable to construct conversion factors for various broad groups of non-traded goods and services. If input–output data are available this can be a source of information on input proportions which can then be adjusted by previously determined conversion factors and summed to give the sector's average conversion factor. This approach can be used, for example, to estimate conversion factors for electricity and transport.

Unskilled labour

In most economic project appraisals, a separate conversion factor is used in estimating the shadow price of unskilled labour. To calculate the conversion factor for unskilled labour we must first determine the opportunity cost of employing labour on a project in terms of the impact on output elsewhere in the economy, e.g. the reduction in agricultural output that results from a movement of labour from farm employment to a new construction project. If the agricultural output is a traded good, e.g. rice, then the conversion factor for the item is applied to the wage rate to give the shadow price of unskilled labour. If the output forgone is diffused, it will be necessary to estimate an average conversion factor as the weighted mean of the conversion factors for each of the outputs (some of which may be non-traded). The average conversion factor is then applied to the market value of the opportunity cost of the unskilled labour employed on the project to give its shadow price value.

Skilled labour is normally valued at the market wage rate, which is converted to a shadow price value using the standard conversion factor. This approach is justified on the grounds that skilled labour is usually in scarce supply and can command an equal wage in other activities, and so the market wage will equal its marginal productivity in alternative employment.

THE SOCIAL DISCOUNT RATE

In the discussion on financial project appraisal we saw how the net cash flow was discounted to give the project's net present value. To do this required a choice as to the rate of discount and in the financial analysis the market rate of interest was used on the grounds that it reflected individuals' rate of time preference.

The same discounting procedure is applied to the stream of shadow net benefits over the life of the project, using a shadow rate of discount which is usually referred to as the social discount rate. The argument for using a shadow rate of discount, rather than the market rate of interest, is the same as for the project's benefits and costs, namely that there are market imperfections or distortions which cause the market interest rate to deviate from its efficiency (shadow) value. For example, the government may operate a policy of controls on interest rate levels, or may allocate investment funds at subsidized rates to certain users. If the market rate of interest is judged to be 'distorted' it is necessary to calculate the opportunity cost of the funds that are to be invested in the project. Since the capital invested in the project could be invested elsewhere in the economy, it is argued that the opportunity cost should be measured by the marginal rate of return on public sector investment. The discount rate will need to be expressed in the same units (numeraire) as the project's benefits and costs. In the Little Mirrlees–World Bank approach, the numeraire is foreign exchange values. The estimate of marginal productivity on public sector investment will therefore need to be converted into shadow price terms using an appropriate conversion factor, possibly the SCF. (The foreign exchange numeraire discount rate is sometimes referred to as the accounting rate of interest (ARI).)

Unfortunately, the application of this approach to calculating the social discount rate is not easy in practice, and the estimated shadow rate of discount will frequently be a rough approximation. It will therefore often be necessary to treat the social discount rate as the best-guess as to its 'true' value and to calculate several NPVs for the project, using a range of discount rate values to test for the sensitivity of the project's NPV to the choice of discount rate.

DISTRIBUTIONAL ISSUES

So far the discussion has assumed that the government's sole objective is to maximize output benefits by achieving an economically efficient allocation of resources. This assumption has allowed us to define shadow prices in efficiency-price terms, and to equate the shadow prices to the (hypothetical) perfectly competitive, Pareto-optimum market equilibrium values.

As we have seen in earlier chapters, the government's objectives need not be limited to economic efficiency, but may include a concern for distributional issues. Can the procedures for project appraisal be extended to incorporate distributional objectives?

The Little Mirrlees–World Bank methodology discusses two distributional concerns. The first relates to the distribution of the economy's resources between present and future consumption. In many low-income economies there are difficulties in raising government revenue through the fiscal system because of the structural underdevelopment of the tax base, and as a consequence the level of public revenue will be lower than the socially desirable level. In such situations, known as having an aggregate savings constraint, it may be appropriate to attach a higher weight to the proportion of a project's output which will go into savings and investment than to the proportion that will be used for immediate consumption.

The second distributional concern relates to the distribution of a project's output among households with different income levels. If the government favours a redistribution of income towards lower-income households, then it will be desirable to attach higher weights to the proportion of the project output accruing to low-income households, and lower weights to the benefits going to higher-income households.

The Little Mirrlees–World Bank approach provides detailed procedures for calculating both types of distributional weights and for incorporating them into the shadow price valuation of the project's net benefits. We shall not describe these methods here, but details can be found in the readings listed at the end of this chapter. There are sizeable problems in applying the distributional weighting method and the approach is seldom used in practice. The distributional weights rely on the planners' value judgements and, in addition, the calculation process involves making a number of assumptions about the likely values of the underlying parameters. Furthermore, the effect of introducing distributional weights can outweigh the efficiency-based shadow prices, making the project selection depend heavily on the arbitrarily chosen distributional weights. For this reason, practitioners have been reluctant to make use of the distribution-weighting approach to project appraisal. A compromise is possible, however, by keeping the two approaches separate, first measuring the NPV at efficiency shadow prices and then identifying separately the redistribution impact of the project. The two sets of estimates can then be compared, allowing the decision maker to make an explicit and transparent choice as to the trade-off between efficiency and distributional objectives.

POLICY ANALYSIS AND COST-BENEFIT ANALYSIS

This chapter has concentrated on the detailed procedures for the appraisal of individual projects using shadow prices. Both CBA and shadow prices have a much broader use, however, and are increasingly recognized as essential techniques for the evaluation of policy changes, and not merely investment projects.

Shadow prices represent the efficiency values that would be generated by perfectly functioning markets. Since price is equal to marginal opportunity cost in this ideal situation, shadow prices can also be defined as the value, measured in terms of the economy's objective(s), of a marginal change in any output or input. Thought of in these terms, it is clear that knowledge of shadow prices will be invaluable information on which to base policy decisions aimed at maximizing national welfare. Little and Mirrlees have recently put the point succinctly:

> It is sometimes argued that the thrust of policy should be to get the prices right, and it is suggested that this is an argument for forgetting about, or at least de-emphasising, shadow prices. To the extent that activities are private, and there is no price or profit control, this is feasible and may be desirable. But in the public sector, or whenever there is public regulation, getting the prices right implies knowing the shadow prices.
>
> (Little and Mirrlees 1990: 363)

This implies a shift in the way in which shadow price information is used. In the late 1960s and 1970s, when the shadow price methodology was being developed and refined, attention was almost entirely confined to the use of shadow prices in project appraisal. Government economic policies were recognized as a source of market distortions, but the distorting policy was regarded as a fixed constraint, and the shadow prices were incorporated in the project appraisal in order to correct for, or offset, these distortions. But an alternative and more direct method of correcting for the inefficiencies caused by market distortions would be to remove the source of the distortion by correcting the policy itself. With the increasing emphasis during the 1980s on economic policy reform, shadow prices came to be used more as an input to economic policy analysis, and economic project appraisal using shadow prices is now used relatively little.

Paradoxically, the recent decline in the application of economic project appraisal methods has resulted in a much wider and more influential use of shadow prices and CBA as tools for development policy analysis.

FURTHER READING

The Little Mirrlees–World Bank methodology of economic project appraisal is developed in Little and Mirrlees (1974) and Squire and van der Tak (1975). A more practical guide to the same approach is contained in Overseas Development Association (1988). Ward and Daren (1991) give numerous examples of the application of the methodology to development projects, and also stress the use of project appraisal as an instrument for policy analysis. Little and Mirrlees (1990) give an interesting assessment of the impact of their ideas on project appraisal practices in the World Bank and other development institutions.

BIBLIOGRAPHY

Adelman, I. (1961) *Theories of Economic Growth and Development*, Stanford: Stanford University Press.

Adelman, I. and Robinson, S. (1978) *Income Distribution Policy in Developing Countries: A Case Study of Korea*, Stanford: Stanford University Press.

—— and —— (1988) 'Macro-economic adjustment and income distribution: alternative models in two economies', *Journal of Development Economics*, 29.

Bandara, J. (1991a) 'Computable general equilibrium models for development policy analysis in LDCs', *Journal of Economic Surveys* (1).

—— (1991b) 'An investigation of "Dutch disease" economics with a miniature computable general equilibrium model', *Journal of Policy Modelling* 13 (1).

Bhagwati, J. and Chakravarty, S. (1969) 'Contributions to Indian economic analysis, a survey' *American Economic Review* 59 (4), Part 2.

Blitzer, C., Clark, P. and Taylor, L. (eds) (1975) *Economy-wide Models and Development Planning*, London: Oxford University Press for the World Bank.

Bulmer-Thomas, V. (1982) *Input–Output Analysis in Developing Countries*, London: Wiley.

Chenery, H. (1955) 'The role of industrialisation in development programs', *American Economic Review, Papers and Proceedings* 45.

—— (1961) 'Comparative advantage and development policy', *American Economic Review* 51 (1).

—— (1972) 'Notes on the use of models in development planning', in M. Faber and D. Seers (eds) *The Crisis in Planning*, London: Chatto & Windus.

Chenery, H. and Bruno, M. (1962) 'Development alternatives in an open economy: the case of Israel', *Economic Journal* 72.

Cline, W. (1972) *Potential Effects of Income Redistribution on Economic Growth; Latin American Cases*, New York: Praeger.

Cohen, J. (1978) *Multi-Objective Programming and Planning*, New York: Academic Press.

123

Cohen, S., Cornelisse, P., Teekens, R. and Thorbecke, E. (eds) (1984) *The Modelling of Socio-Economic Planning Processes*, London: Gower.

Dasgupta, P.S., Marglin, S.A. and Sen, A.K. (1972) *Guidelines for Project Evaluation*, Vienna: United Nations (UNIDO).

Dervis, K., de Melo, J. and Robinson, S. (eds) (1982) *General Equilibrium Models for Developing Countries*, London: Cambridge University Press.

Dinwiddly, C. and Teal, F. (1988) *The Two-Sector General Equilibrium Model: A New Approach*, London: Philip Allan/St Martin's Press.

Domar, E. (1957) 'A Soviet model of growth', in E. Domar (ed.) *Essays in the Theory of Economic Growth*, Cambridge: Cambridge University Press.

Dorfman, R. (1953) 'Mathematical or linear programming: a non-mathematical exposition', *American Economic Review* 43 (5).

Dorfman, R., Samuelson, P. and Solow, R. (1958) *Linear Programming and Economics Analysis*, New York: McGraw-Hill.

Erlich, A. (1960) *The Soviet Industrialisation Debate, 1924–28*, Cambridge: Cambridge University Press.

Faber, M. and Seers, D. (eds) (1972) *The Crisis in Planning*, London: Chatto & Windus.

Fox, K., Sengupta, J. and Thorbecke, E. (1966) *The Theory of Quantitative Economic Policy with Application to Economic Growth and Stabilisation*, New York: North-Holland.

Frisch, R. (1958) *A Method of Working Out a Macro-economic Plan Frame with Particular Reference to the Evaluation of Development Projects, Foreign Trade and Employment*, Oslo: Oslo University Press.

Gillis, M., Perkins, D.H., Roemer, M. and Snodgrass, D.R. (1992) *Economics of Development*, 3rd edn, New York: W.W. Norton.

Government of India (1973) *A Technical Note on the Approach to the Fifth Five-Year Plan of India*, New Delhi: Planning Commission.

Gurley, J. and Shaw, E. (1955) 'Financial aspects of economic development', *American Economic Review* 45 (4).

Hirschman, A. (1958) *The Strategy of Economic Development*, New Haven: Yale University Press.

Islam, N. (1970) 'The relevance of development models to economic planning in developing countries', *Economic Bulletin for Asia and the Far East*, June–September.

Killick, T. (1976) 'The possibilities of development planning', *Oxford Economic Papers*, July.

Kornai, J. (1970) 'A general descriptive model of planning processes', *Economics of Planning* 10 (1).

—— (1975) 'Models and policy: the dialogue between model builder and planner', in C. Blitzer, P. Clark and L. Taylor (eds) *Economy-wide Models and Development Planning*, London: Oxford University Press for the World Bank.

Little, I.M.D. (1982) *Economic Development: Theory, Policy and International Relations*, New York: Basic Books.

Little, I. M. D. and Mirrlees, J. (1974) *Project Appraisal and Planning for Developing Countries*, London: Heinemann.

—— and —— (1990) 'Project appraisal and planning twenty years on', *Proceedings of the World Bank Annual Conference on Development Economics*, Washington, DC: World Bank.

Mahalanobis, P. (1953) 'Some observations on the process of growth of national income', *Saikhya* 12.

Manne, A. (1974) 'Multi-sectoral models for development planning: a survey', *Journal of Development Economics* 1 (1).

Mansur, A. and Khondker, B. (1991) 'Revenue effects of the VAT system', *Bangladesh Development Studies* 19 (3).

Maton, J. and Joos, M. (1984) 'A simplified dynamic SAM model of Malaysia: the effects of technical progress, capital accumulation and income distribution on employment', in S. Cohen, P. Cornelisse, R. Teekens and E. Thorbecke (eds) *The Modelling of Socio-Economic Planning Processes*, London: Gower.

McKinnon, R. (1973) *Money and Capital in Economic Development*, Washington, DC: Brookings Institution.

Myrdal, G. (1968) *Asian Drama*, vol. III, London: Penguin Books.

Overseas Development Administration (1988) *Appraisal of Projects in Developing Countries*, London: Her Majesty's Stationery Office.

Pyatt, G. and Thorbecke, E. (1976) *Planning Technique for a Better Future*, Geneva: ILO.

Pyatt, G., Roe, A. and Stone, R. (1977) *Social Accounting Matrices for Development Planning with Special Reference to Sri Lanka*, Cambridge: Cambridge University Press.

Quaddus, M. and Chowdhury, A. (1990) 'Social preference function and policy prioritisation for Bangladesh: an experiment with analytical hierarchy process', *Economics of Planning* 23.

Rao, V. K. R. B. (1952) 'Investment, income and multiplier in an underdeveloped economy', *Indian Economy Review*, February.

Rao, V. V. B. (1988) 'Measurement of deprivation and poverty based on proportion spent on food: an exploratory exercise', *World Development* 9 (4).

Robinson, S. (1989) 'Multi-sectoral models', in H. Chenery and T. Srinivasan (eds) *Handbook of Development Economics*, New York: North-Holland.

—— (1991) 'Macro-economics, financial variables and computable general equilibrium models', *World Development* 19 (11).

Robinson, S. and Tyson, L. (1984) 'Modelling structural adjustment: micro and macro elements in a general equilibrium framework', in H. Scarf and J. Shoven (eds) *Applied General Equilibrium Analysis*, Cambridge: Cambridge University Press.

Rosenstein-Rodan, P. (1943) 'Problems of industrialisation of Eastern and South-Eastern Europe', *Economic Journal*, June–September.

Rostow, W. W. (1960) *The Stages of Economic Growth*, Cambridge, MA: MIT Press.

Sen, A. K. (1972) 'Accounting price and control areas: an approach to project evaluation', *Economic Journal* 82.

Squire, L. and van der Tak, H. G. (1975) *Economic Analysis of Projects*, Baltimore: Johns Hopkins University Press for the World Bank.

Taylor, L. (1979) *Macro Models for Developing Countries*, New York: McGraw-Hill.
—— (ed.) (1990) *Socially Relevant Policy Analysis*, Cambridge, MA: MIT Press.
Thirlwall, A. (1983) *Growth and Development*, 3rd edn, London: Macmillan.
Tinbergen, J. (1955) 'The relevance of theoretical criteria in the selection of investment planning', in M. Millikan (ed.) *Investment Criteria and Economic Growth*, Cambridge: Cambridge University Press.
—— (1958) *The Design of Development*, Baltimore: Johns Hopkins University Press.
—— (1981) 'Nobel Lecture', reprinted in *American Economic Review*, December.
Tinbergen, J. and Bos, H. (1962) *Mathematical Models of Economic Growth*, New York: McGraw-Hill.
Todaro, M. (1971) *Development Planning*, Nairobi: Oxford University Press.
—— (1989) *Economic Development in the Third World*, 4th edn, Harlow: Longman.
Ward, W. A. and Daren, B. J. (1991) *The Economics of Project Analysis*, Washington, DC: Economic Development Institute of the World Bank.
Weber, J. E. (1982) *Mathematical Analysis*, New York: Harper & Row.

INDEX

Note: Page numbers in **bold** type refer to **figures** and page numbers in *italic* type refer to *tables*.

929010